"十四五"职业教育部委级

教育部国家职业教育专业教学资源民族文化传承与创新子库
"中国丝绸技艺民族文化传承与创新"配套双语教材
江苏省高等职业院校高水平专业群"纺织品检验与贸易"配套教材

丝绸旗袍文化
Silk Cheongsam Culture

陆 洁　王海燕◎主　编

郑蓉蓉◎副主编

翟　芳◎译

中国纺织出版社有限公司

内容提要 /Summary

本书根据旗袍发展的自然脉络、旗袍成型的步骤，从旗袍的起源演变、文化内涵、传承现状、分类流派、设计理念、款式结构、制作工艺、面料图案和装饰细节等方面，深入介绍了与旗袍相关的知识。使学生在阅读与实践的过程中逐渐掌握旗袍成型的基本流程与要求，体会旗袍作为传统文化与现代观念结合的典型例证所呈现出的现实意义与历史印记。

This book intends to give a full picture of cheongsam (Qipao in Chinese), which covers the evolution, culture, inheritance, classification, design concepts, styles, cheongsam-making techniques, fabric patterns and decoration of cheongsam in line with its development process. The purpose of this book is to enable students to understand the basic process and requirements for cheongsam formation step by step through reading and practice, and to understand the practical and historical significance of the dress as a typical example of the combination of traditional culture and modern concepts.

图书在版编目（CIP）数据

丝绸旗袍文化 = Silk Cheongsam Culture：汉、英 / 陆洁，王海燕主编；郑蓉蓉副主编；翟芳译. -- 北京：中国纺织出版社有限公司，2025．4．--（"十四五"职业教育部委级规划教材）（教育部国家职业教育专业教学资源民族文化传承与创新子库"中国丝绸技艺民族文化传承与创新"配套双语教材）（江苏省高等职业院校高水平专业群"纺织品检验与贸易"配套教材）. -- ISBN 978-7-5229-1967-6

Ⅰ．TS941.717.8

中国国家版本馆 CIP 数据核字第 20244DK415 号

责任编辑：孔会云　朱利锋　责任校对：高　涵　责任印制：王艳丽

中国纺织出版社有限公司出版发行
地址：北京市朝阳区百子湾东里 A407 号楼　邮政编码：100124
销售电话：010—67004422　传真：010—87155801
http://www.c-textilep.com
中国纺织出版社天猫旗舰店
官方微博 http://weibo.com/2119887771
北京通天印刷有限责任公司印刷　各地新华书店经销
2025 年 4 月第 1 版第 1 次印刷
开本：787×1092　1/16　印张：8.5
字数：175 千字　定价：88.00 元

前 言/**Foreword**

　　旗袍是中华民族的优秀传统文化之一，了解并传承旗袍技艺精髓，不仅有助于丰富学生的服装专业知识，更能帮助学生建立正确的辩证观念，全面提高传统文化素养，增强我国文化软实力。本书图文并茂地介绍了旗袍各个阶段的变化与创新，结合当前研究前沿的观念动态，希望帮助学生从民族与时代相结合的视角获取一定的艺术创作灵感，使学生在阅读与实践的过程中建立起将传统文化与现代设计理念有机结合的纽带，并从中找到行之有效的灵感激发源泉与创作表现方法。

　　Understanding and passing on the essential techniques of cheongsam, one of the remarkable traditional cultures of the Chinese nation, can not only help students enrich their professional knowledge of clothing, but also enable students to establish dialectical concepts and strengthen their insights into traditional culture comprehensively to boost China's cultural soft power. By introducing the changes and innovations of cheongsam at each stage with texts and illustrations, the book aims at helping students gain artistic creation inspiration from the perspective of the combination of the nation and the times in the context of current cutting-edge research trends, thus allowing them to build a link between traditional culture and modern design concepts amid reading and practice, and discover effective sources of inspiration and expression methods.

　　旗袍随着时代变迁不断更新的历程深受政治、经济、思想文化发展的复杂影响。旗袍的诞生离不开五四运动和新文化运动的启蒙教育，女性身体与思想的解放在旗袍的款式结构上体现得非常明显，从传统平面的隐藏式外形到现代立体的表露式构成，服装之于人体表面及内在的作用被彻底地否定性重塑；旗袍的演变过程也不断受到西方科技与文艺思潮、流行趋势的影响，从面料配饰到穿着方式的转变，使旗袍及其穿着者备受国际社会的普遍关注，女性主权的确立作为社会进步的标志，也成为与服装演变密切相关的重要依据。总之，旗袍是一个时代的产物，反观其对人类社会的影响，也是深远而意义重大的；服装作为文化的组成部分，无论从人类学还是社会学的角度，以及艺术设计本身而言，都将一路伴随并反映穿着者的思想行为。因此，学习旗袍文化的相关内容是一种以个例见群体的研究方法，诚愿读者能够从中窥见引以为鉴的灵感源头，作为创意设计的崭新起点。

　　Cheongsams have been constantly innovated with the changing times, which reflects the complex influence of political, economic, ideological and cultural development. The birth of cheongsam is inseparable from the enlightenment of the May Fourth Movement and the New Culture Movement. The liberation of women's bodies and minds is the most evident in the style and structure of cheongsam which changed from the traditional flat cut to the modern three-dimensional structure, leading to the thorough redefinition of the function of clothing for human body and spirit. The Western technology, coupled with its literary and artistic development and fashion trend, also contributed to cheongsam's evolution. Due to the changing style and accessories of the dress, cheongsam and its wearers have attracted much attention from the international community. Moreover, the establishment of female sovereignty, a symbol of social progress, becomes another significant basis with close relation to clothing's evolution. In short, cheongsam, as a product of a certain era, exerts a far-reaching and profound impact on human society. As an integral part of culture, clothing acts as a reflection of the thought and behavior of its wearers both in anthropology and sociology. For this reason, studying cheongsam culture is a case study approach to learn more about clothing culture. I sincerely hope that readers can get inspirations for creative design from this book.

编者 /Editor

2024年4月 /April 2024

译者序
Translator's Preface

《丝绸旗袍文化》是教育部职业教育专业教学资源国家级子库"中华丝织技艺与民族文化传承创新"双语系列教材之一，也可作为对丝绸与服饰文化感兴趣的读者及纺织服装行业从业者的参考书籍。同时，英文版课程的推出，将吸引全球范围内众多该行业从业者的关注，对中国纺织服装行业的国际化进程具有重要意义。

Silk Cheongsam Culture is one of the Bilingual teaching materials "Inheritance and Innovation of Chinese Silk Skills and National Culture", a sub-library of national vocational education professional teaching resources of the Ministry of Education. It can also serve as a reference for readers interested in textiles and clothing as well as practitioners in the textile and clothing industry. The English version of the course will be of interest to many practitioners in the industry both at home and abroad. It will play an important role in the international expansion of the Chinese textile and clothing industry.

全书共分为四章，英文翻译及校对工作由西安工程大学的翟芳老师担任，西安工程大学人文社会科学学院胡伟华教授进行了全程指导和专业审校，提出了诸多建设性的修改意见，确保了译文的准确性和可读性。

This book consists of four chapters, all the four chapters were translated and proofread by Zhai Fang at Xi'an Polytechnic University. Dr. Weihua Hu, a professor at the School of Humanities and Social Sciences of Xi'an Polytechnic University, helped to review all the translations and put forward valuable suggestions for revision.

教学内容及课时安排
Teaching Content and Schedule

章/课时	课程性质/课时	节	课程内容
第一章 旗袍文化之源 （6课时）	理论（6课时）	一	第一节　旗袍起源
		二	第二节　旗袍成型
		三	第三节　旗袍演变
第二章 旗袍造型之思 （18课时）	理论（6课时） 实践（12课时）	一	第一节　旗袍款式设计
		二	第二节　旗袍结构设计
第三章 旗袍匠艺之心 （18课时）	理论（6课时） 实践（12课时）	一	第一节　旗袍排料裁剪
		二	第二节　旗袍制作工艺
第四章 旗袍材质之美 （12课时）	理论（6课时） 实践（6课时）	一	第一节　旗袍与丝绸面料
		二	第二节　丝绸旗袍图案风格
		三	第三节　丝绸旗袍装饰手法

注　各院校可根据自身的教学特点和教学计划对课程时数进行调整。

Chapter/Credit Hours	Course Nature/Credit Hours	Class	Course Content
Chapter One: The Origin of Cheongsam Culture (6 Credit Hours)	Theory (6 Credit Hours)	Section 1	1.1　The Origin of Cheongsam
		Section 2	1.2　The Forming of Cheongsam
		Section 3	1.3　The Evolution of Cheongsam
Chapter Two: Thoughts on the Shape of Cheongsam (18 Credit Hours)	Theory (6 Credit Hours) Practice (12 Credit Hours)	Section 1	2.1　Cheongsam Style Design
		Section 2	2.2　Cheongsam Structure Design
Chapter Three: The Heart of Cheongsam Craftsmanship (18 Credit Hours)	Theory (6 Credit Hours) Practice (12 Credit Hours)	Section 1	3.1　Cheongsam Pattern Cutting
		Section 2	3.2　Cheongsam Production Process
Chapter Four: The Beauty of Cheongsam Materials (12 Credit Hours)	Theory (6 Credit Hours) Practice (6 Credit Hours)	Section 1	4.1　Cheongsam and Silk Fabric
		Section 2	4.2　Pattern Style of Silk Cheongsam
		Section 3	4.3　Decorative Techniques of Silk Cheongsam

Note　The credit hours can be adjusted according to the institution's teaching characteristics and teaching plan.

目 录/Contents

○ 第一章 /
旗袍文化之源
The Origin of Cheongsam Culture

第一节　旗袍起源/The Origin of Cheongsam

关于旗袍的起源，学术界有很多争论，主要观点可概括为以下四种。

There is a lot of debate in the academic circle regarding the origin of cheongsam, and the main views can be summarized into four groups as follows.

第一种以周锡保先生的《中国古代服饰史》为代表，认为旗袍是从清代旗女的袍服直接发展而来。但也有学者认为：民国旗袍虽然具有类似于旗女之袍的形式，却不再具有旗女之袍的含义，如果完全认为民国旗袍是直接由旗女之袍发展而来，未免有失偏颇。

The first group is represented by the *History of Ancient Chinese Costumes*, a work of Mr. Zhou Xibao, which believed that cheongsam is directly developed from the robe of the banner women in the Qing Dynasty. However, some scholars considered such a point of view a bit biased, since the cheongsam that appeared in the Republic of China was similar to the robe of the banner women in the form but different in the connotation.

第二种以袁杰英教授的《中国旗袍》及包铭新教授的《中国旗袍》《近代中国女装实录》为代表，认为旗袍和旗装袍有一定继承关系，同时认为，旗袍的源头应是西周麻布窄形筒装或先秦两汉的深衣。江南大学崔荣荣教授也在著作《近代汉族民间服饰全集》中写道："有些人质疑旗袍是满族服饰……我认为这些认知是片面和表面化的，汉族的袍服已有2000多年的历史，而满族的袍和褂的历史渊源又从哪里来的呢？答案我想很容易得到。"正如国学大师章太炎所言："昔诸葛亮造筒袖铠……满洲之服，其筒袖铠之绪也。"将满族服饰的源头追溯到三国时期的蜀汉服饰。

The second group, represented by Professor Yuan Jieying's *Chinese Cheongsam* and Professor Bao Mingxin's *Chinese Cheongsam* and *Records of Modern Chinese Women's Wear*, believed that there is a certain inheritance between the cheongsam and the robe of the banner women, yet its origin should be the narrow straight linen cloth of the Western Zhou Dynasty or the Shen Clothing of the pre-Qin and Han Dynasties. In his book *A Collection of Contemporary Han Folk Costumes*,

Professor Cui Rongrong of Jiangnan University pointed out: "It's lopsided and superficial to doubt that cheongsam belongs to Manchu clothing. Han robes have a history of more than 2, 000 years, where do the historical origins of the Manchu robes and gowns come from? I think the answer is evident." As Zhang Taiyan, a master of Chinese studies, stated: " The sleeve armor made by Zhuge Liang (a famous politician in the Three Kingdoms Period) is the origin of the Manchu clothing." These opinions regarded that the origin of Manchu costumes can be traced back to the costumes of the Kingdom of Shu Han (221–263) in the Three Kingdoms Period.

第三种以王宇清的《历代妇女袍服考实》为代表，认为中国妇女所穿的袍可远溯周、秦、汉、唐、宋、明时代，并不是只有在清代旗女才穿袍服。他认为旗女之袍对民国旗袍有一定影响，但并不认为二者有直接继承关系，因此认为民国旗袍称为"旗"袍并不合适，他倡导将旗袍改名为"祺袍"。

The third group, represented by Wang Yuqing's *The Research and Examination of Chinese Women's Gown in Successive Dynasties*, held that robes worn by Chinese women can date back to the Zhou, Qin, Han, Tang, Song and Ming Dynasties, not only in the Qing Dynasty. He agreed that the cheongsam in the Republic of China was influenced by the robe of the banner women, but denied a direct inheritance between the two. For this reason, he proposed that the cheongsam should be renamed "Qi Pao (祺袍)" rather than "Qi Pao (旗袍)".

第四种以下向阳教授的《论旗袍的流行起源》为代表，认为旗袍是中国服装传统的西化变异。融合了旗袍马甲和文明新装的特点，同时又结合了西式裙装的配伍形式，构成了既具有西方流行的影子，又具有鲜明中国特色和时代象征的新时尚流行和服装审美的特点，可视为中西服饰交融的设计典范。因此他认为，"旗袍"这个名称其实是对这类女装的一种"误称"，因为早期倡导旗袍的群体主要是都市中受西学影响较深、追求男女平等、反对封建礼教的新女性等社会群体，她（他）们绝大多数是汉族人，她（他）们的祖先在清初经过流血抗争才为汉族妇女取得不穿满式服装的权利，她们不可能去复辟帝制时代的异族压迫者的服饰，所以旗装袍在民国复辟的条件并不具备。

The fourth group, taking Professor Bian Xiangyang's *On the Genesis of the Prevalence of Qi Pao* as the representative, deemed that the cheongsam is a variation of traditional Chinese clothing under the Western influence. It integrates the characteristics of the cheongsam vest and the new civilized clothes, as well as the matching of Western-style skirts, forming a new clothing fashion and a symbol of the times influenced by the Western fashion with distinctive Chinese characteristics. It can be regarded as a design model for the fusion of Chinese and Western clothing. Hence, he thought "cheongsam" is actually a "misnomer" for this type of women's clothing, because the groups who advocated such clothing in earlier times were the new women and students in cities who were deeply influenced by Western thoughts, pursued equality between men and women, and opposed feudal ethics, and a majority of them were Han people. Her (his) ancestors obtained the right not to wear Manchu clothing for Han women through bloody struggles in the early Qing Dynasty. Therefore, it's not likely for them to restore the costumes of the imperial system established by the alien oppressors. In other words, the conditions for the restoration of the cheongsam in the Republic of China were not met.

综合以上几种观点，本书作者认为，旗袍属于袍服的范畴，在其诞生之初是一种上下连属、前后连裁的平面结构形态的服装。旗袍不是在民国建立时出现的，而是在潮流演变过程中逐渐出现的一种类似旗装的"长马甲"，平直宽大，衣长及足，内穿喇叭形宽大袖子的短袄，后来二者合一，形成民国旗袍的雏形。

Based on the above viewpoints, the author of this book believes that the cheongsam falls into the category of robes, and at the beginning of its birth, it was a kind of one-piece dress with a flat structure that is cut from front to back. Additionally, the cheongsam did not appear when the Republic of China was established, but gradually transform into a kind of straight, loose, floor-length "long vest" similar to the robe of banner people, which evolved to be the prototype of the cheongsam in the Republic of China by integrating with the short jacket with flared wide sleeves inside.

一、旗袍萌芽/The Generation of Cheongsam

（一）男女袍服/Men's and Women's Gowns

1. 女子袍服/Women's Gowns

清末满族女子的常服一般称为"衬衣"和"氅衣"。如果从"旗女之袍"着手分析，旗袍的源头似乎并不在同为两侧开衩的旗装氅衣里，因为氅衣不能单穿，必须搭配衬衣穿着。氅衣多在两侧开衩处做一个特别的装饰，称"如意云头"。而旗袍诞生之初有大量无衩或开小衩的式样，但这些特征在旗袍定型之后几乎消失殆尽。所以，即便旗袍的源头在旗装，也应该是无衩的衬衣类服饰，才更符合旗袍本身的演变逻辑，如图1-1所示。

In the late Qing Dynasty, the informal clothing of Manchu women consisted of "under robe" and "outer garment (Chang Yi)". If it is analyzed based on banner women's gown, the origin of cheongsam does not seem to be the outer garment of the banner gown with slits on both sides, because such gown can not be worn alone, it must be matched with an under robe. The outer garment is often decorated with something special called "Ruyi Yun Tou (a pattern that symbolizes good luck)" at the slits on both sides. In the early days upon the birth of cheongsam, there were a lot of styles without opening or with small openings, but these features virtually disappeared when the cheongsam was finalized. Therefore, even if the cheongsam originated from the banner gown, it should be the under robe without openings, which is more in accord with its evolution logic, as shown in Figure 1-1.

立领和大襟对于现代人来说，是极具特色的传统视觉元素。然而放在当时的社会背景里却很普遍，这些元素是满、汉服饰中常用的基本样式，而不是区分这两类服饰的必要条件，可见当时旗装与汉装早已渐趋融合。立领一直是汉装的元素，清代末期才逐步用在旗装

图 1-1　氅衣与衬衣的区别
Difference between the Outer Garment and the Under Robe

上，加上繁复的装饰和宽松的衣身，这种新式旗装与满人之前的服装已相距甚远，如图 1-2 所示。

For modern people, the mandarin collar and the front of a garment represent distinctive traditional visual elements. However, they were very common in Qing Dynasty as the basic style of both the Manchu and Hanfu（Han-style clothing）, rather than a necessary condition for distinguishing between these two types of costumes. It can be seen from this fact that the two styles of clothing were gradually integrated at that time. As a typical element of Hanfu of all time, the mandarin collar was used in the banner gown in the late Qing Dynasty, which coupled with complicated decorations and loose clothes and made the banner gown quite different from the costume that the Manchus wore before, as shown in Figure 1-2.

图 1-2　清末期与清早期的服饰对比
Comparison of Clothing in the Late Qing Dynasty and the Early Qing Dynasty

2. 男子袍服/Men's Gowns

自汉代后，中国汉族女性服饰逐渐转变为"上衣下裳"式，俗称"两截衣"，穿袍服几乎成为男性的专利。女性穿"两截衣"成了封建礼教对女性压迫的象征，《女儿经》中就写道："为甚事，两截衣，女人不与丈夫齐。百凡事体须卑顺，不得司晨啼母鸡。"

Since the Han Dynasty, Chinese Han women gradually only wore the upper and the lower garments, commonly known as "Liang Jie Yi（two separated clothing）", while the outer garment has become exclusive to men. Therefore, "Liang Jie Yi has become a symbol of the oppression of women by feudal ethics. In the *Nu Er Lapelg*（Three–character Lines of Advice and Admonitions for Young Women of every household）, it is written that："Women should wear the upper and the lower garments to do farm work, and can not share equal rights with their husbands. They are required to be obedient to their husbands."

20世纪初，中西方女性已不满足于仅仅留守家中。受社会变革和进步思想的影响，女性逐渐意识到自己不仅有家庭的角色，还应该是社会的一员。她们可以通过接受知识和技能的培训，建立独立自主的社会地位，摆脱对男性的依赖。中国经过辛亥革命的洗礼，整个社会洋溢着为民族崛起而奋斗的气息，女性则一直积极谋求与男性平等的权利，包括受教育的权利、外出的权利，以及参与民族复兴的权利。随着反对缠足、兴办女学等运动的推进，职业女性大量涌现，女性争取自由平等的权利得到了初步实现，许多勇于突破传统的女子，纷纷以穿男装、梳辫子为新潮，如图1–3所示，她们以此为标志显示自己不受封建思想的束缚。尤其是那些坚信共和、民主等进步思想的女性，她们认为传统袄裙的上下分穿方式不如男子的一袭袍服更代表革命的彻底性。张爱玲在《更衣记》中就曾这样写道："五族共和之后，全国妇女突然一致采用旗袍，倒不是为了效忠于清朝提倡复辟运动，而是因为女子蓄意要模仿男子。她们初受西方文化的熏陶，醉心于男女平权之说，可是四周的情形与理想相差太远了，羞愤之下，她们排斥女性化的一切，恨不得将女人的根性斩尽杀绝。"

In the early 20th century, both Chinese and Western women were no longer satisfied with the status of staying at home. Influenced by social changes and progressive ideas, women realized that their roles should not be restricted to family only, but shall be extended to the society, that is, they can establish independent social status to get rid of dependence on men by receiving trainings on knowledge and skills. Benefiting from the Revolution of 1911, the faith in striving for the rise of the nation had been spread everywhere in China. In such context, women made ongoing efforts to seek equal rights with men, including the right of education, the right of going out for work, and the right to be responsible for national rejuvenation. With the advancement of the movement against foot–binding and the establishment of women's schools, a large number of professional women stood out, and women's rights to fight for freedom and equality achieved initial success. Many women with the courage to break away from tradition had adopted men's clothing and braids as new trends, as shown in Figure 1–3, and they took it as a sign to show their objection to feudal ideology. Particularly the women with strong faith in progressive ideas such as republic and democracy thought that wearing the upper and lower garments can not demonstrate their strong belief in the revolutionary spirit as men who wore the one-piece robes. Eileen Chang, a famous Chinese writer, in her work *Chinese Life and Fashions*, wrote："After the Republic of Five Nationalities, women all over the country suddenly wore cheongsams, not because they were loyal to the Qing Dynasty and advocated the restoration movement, but because they deliberately wanted to imitate men. As the early receivers of the western culture, women in this times were rapt in the idea of equal rights between men and women, yet the reality they faced was too far from the ideal. Driven by such shame and anger, they rejected everything that is feminine, even wanted to eradicate the nature of women."

图1-3 穿男装的女子
A Woman in Men's Clothing

（二）倒大袖短袄/Big Sleeve Short Jacket

清朝汉族妇女通常穿两截式的衣裙，俗称"裙褂"或"裤褂"，这种穿着方式一直延续到清末民初，如图1-4所示。1911年辛亥革命爆发，推翻了中国历史上最后一个封建王朝，为西式服装在中国的普及清除了政治障碍，同时也摒弃了传统苛刻的礼教与风化观念，打破了服制上等级森严的种种桎梏。服装逐渐走向平民化、国际化的自由变革，逐渐摆脱了沉重的传统束缚。服装装饰一扫清朝的矫饰之风，趋向于简洁，色调力求淡雅，注重展现女性的自然之美，被称为"文明新装"，如图1-5所示。上衣仍采用整片式裁剪，胸围和腰围和下摆处则逐渐趋向合体设计。

In the Qing Dynasty, Han women usually wore the upper and the lower garments, commonly known as "Qungua" or "Kugua" till the late Qing Dynasty and the early Republic of China, as shown in Figure 1-4. The Revolution of 1911 broke out suddenly and overthrew the last feudal dynasty in Chinese history, clearing the political obstacles for the popularization of Western-style clothing in China, abandoning the traditional and harsh concepts of etiquette and formalities, and reliving the clothing culture from the shackles of strict hierarchy. From then on, clothing culture evolved to be more common and international, gradually unloading the heavy traditional burden. Getting rid of the complicated decoration in the Qing Dynasty, it became simpler with elegant colors, focusing on underlining the natural beauty of women, which was called the "Civilized New Outfit", as shown in Figure 1-5. The upper garment remained to be cut in one piece, the waistline and hem changed to be well-fitted.

图1-4　晚清与清末汉女服装形象
The Image of Han Women's Clothing in the Late Qing Dynasty

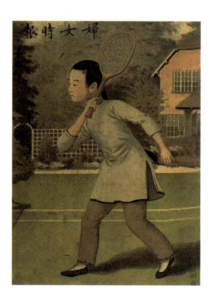 

图1-5　20世纪初的新女性形象
Images of New Women in the Early 20th Century

7

开始时，衣长过臀，两侧开衩，无论袖子还是裤子都设计得十分紧窄。20世纪20年代初，衣长逐渐缩短，下摆改为圆弧形，与喇叭状的袖口构成流畅的外形轮廓，袖子长度达到肘部，并略施纹样作为点缀。下裙长度逐渐缩短，裙褶量逐渐减少，由旧时的围系式褶裙简化为套穿式喇叭裙。

In the beginning, the clothes were longer than the hips, with slits on the sides and very tight sleeves and trousers. In the early 1920s, the length of the dress was gradually shortened, and the hem was changed to an arc shape, which formed a smooth outline with the flared cuffs. The sleeve length reached the elbow and was decorated with simple patterns. The length of the lower garment was shortened with reduced folds, accompanied by a simplified wearing method from the old pleated skirt to a slip-on flared skirt.

随着满族统治政权的消亡，旧式的旗女长袍被摒弃，东、西式装扮熙熙攘攘地共同出现在人们的视野。当时的女学生作为知识女性的代表，成为社会的理想形象，她们是文明的象征、时尚的先导，如图 1-6 所示。这一时期最为流行的倒大袖短袄常与马甲搭配穿着，样式有对襟、琵琶襟、一字襟、大襟、直襟、斜襟等变化，领、袖、襟、摆多镶滚花边或刺绣纹样，衣摆有方有圆，宽窄长短的变化也较多。上衣与裙子的面料除了常用的丝质锦缎外，还有蕾丝、荷叶边、流苏等元素，展现出浓厚的西方色彩，可谓花团锦簇、争奇斗艳，如图 1-7 和图 1-8 所示。

With the demise of the Manchu ruling regime, the old-fashioned banner gowns were abandoned, and the Eastern and Western-style dresses got integrated. The female students at that time, as representatives of educated women, became the ideal image of the society as the symbols of civilization and the forerunners of fashion, as shown in Figure 1-6. In this period, the most popular big sleeved short jackets were often worn with vests. And the style of Lapel (the front of garments) varied from double lapel, Pipa lapel, Yizi lapel, big lapel, straight lapel, slanting lapel, etc. The collar, sleeves, lapel, and hem were decorated with laciness and embroidery patterns. The hem was square and round, varying from width and length. In addition to the commonly used silk brocade, the fabrics of upper garments and skirts also involved lace, ruffles, tassels, etc. The decorations featured prominently in Western style, as shown in Figure 1-7 and Figure 1-8.

图 1-6　培华女子中学校服
Peihua Girls' High School Uniform

图1-7 倒大袖短袄
Big Sleeved Short Jacket

图1-8 倒大袖短袄搭配马甲穿着
Big Sleeved Short Jacket with Vest

二、旗袍雏形/The Prototype of Cheongsam

（一）长马甲旗袍/Long Vest Cheongsam

1.长马甲/Long Vest

长马甲是一种宽松无袖的长衣，搭配当时的倒大袖短袄穿着。穿着时看上去比大圆角短袄更显婀娜修长，是旗袍诞生过程中出现的一个过渡产物，如图1-9所示。

The long vest was a loose, sleeveless long garment worn with the big sleeved short jacket at the time. When being worn, it looked more graceful and slender than the big rounded short jacket. It was a transitional product during the birth of the cheongsam, as shown in Figure 1-9.

图1-9　穿在倒大袖短袄外面的长马甲
A Long Vest Worn over a Big Sleeved Short Jacket

2. 假两件/Fake Two Pieces

将长马甲与短袄缝制在一起，便形成"假两件"，如图1-10所示。在旗袍演变的过程中，长马甲起到了引领潮流的关键作用，将它套穿在短袄外面，既省去了原来穿在短袄下的裙子，又可以遮挡短袄侧面露出的腰部，长马甲与短袄融合后便成为有接袖设计的"假两件"。

When the long vest and the short jacket were sewed together, the "fake two pieces" came into being, as shown in Figure 1-10. Amid the evolution of cheongsam, the long vest played a key role in leading the trend. Wearing it over the short jacket not only hid the skirt originally worn under the short jacket, but also covered the waist exposed on the side of the short jacket. The fusion of the long vest and the short jacket became a "fake two pieces" with a sleeve design.

图1-10　有接袖设计的"假两件"
"Fake Two Pieces" with Sleeve Design

（二）倒大袖旗袍 /Big Sleeve Cheongsam

再后来，"假两件"这种多此一举的做法也被舍弃，出现了倒大袖旗袍。与此同时，一直套穿在袄裙外面的短马甲也十分流行，于是袄裙与长、短马甲，倒大袖旗袍在这一时期同时存在。无衩筒状倒大袖旗袍虽然十分宽松，但显得清丽可人，俏皮大方，如图 1-11 所示。加之外来设计元素的融入，旗袍种类百花齐放，致使旗袍初见时的 20 世纪 20 年代早期远比旗袍定型之后来得更加缤纷有趣。

Later, this superfluous practice "fake two piece" was also abandoned, and the cheongsam with big sleeves emerged. Meanwhile, short vests that have been worn over the jacket were also very popular. In this context, jacket skirts（Aoqun）, long and short vests, and cheongsam with large sleeves emerged in the same era. The straight, big sleeved cheongsam without opening was very loose but beautiful and nifty, as shown in Figure 1-11. With the integration of foreign design elements, the early 1920s when cheongsam first appeared was far more colorful than that after the cheongsam was finalized.

图 1-11　无衩筒状倒大袖旗袍
Straight Big Sleeved Cheongsam without Openings

第二节　旗袍成型/The Forming of Cheongsam

20 世纪 30 年代，影响后来近百年中国女性服装的"旗袍"成了绝对主流。这个年代开始盛行长至脚面的旗袍，曳地旗袍虽不如想象中那么紧身，却散发一种修长纤细的韵味，高领长旗袍始终代表着旗袍岁月里最为亭亭玉立的身影。40 年代后，无袖旗袍独领风骚，面料、装饰与穿着方式更加中西合璧。

In the 1930s, the "Cheongsam" that had a great impact on Chinese women's clothing for nearly a hundred years later became the mainstream. And the floor-length long cheongsam started to get popular in this era. The floor-length cheongsam, though not as tight as we imagined, looked slender. The long cheongsam with a high collar has always been the most typical representative of a slim figure in the cheongsam times. After the 1940s, the sleeveless cheongsam took a leading role, and the Chinese and Western clothing cultures got increasingly intertwined in its fabrics, decorations and wearing methods.

一、曳地旗袍/Floor Length Cheongsam

（一）旗袍款式/Style of Cheongsam

1. 旗袍开衩/Slits of Cheongsam

当时的旗袍仍采用平面剪裁，并不强调S曲线，虽搭配高跟鞋穿着，但没有想象中那么紧身，展现出一种修长纤细的视觉效果，相较20世纪20年代的宽松倒大袖旗袍，它显得更

图1-12　低开衩旗袍
Cheongsam with Low Slits

加苗条。因为旗袍的长度和合体度有所提升，所以侧面的扣子数量也随之增多，却做得并不那么夸张。从侧面和坐姿可以看出，开衩在小腿的地方，隐约可见蕾丝边的浅色衬裙，行走时小腿朦胧而富有层次，如图1-12所示。

At that time, the cheongsam was still flat-cut with little attention paid to the S-curve. Although it was worn with high-heeled shoes, the floor-length cheongsam was not as tight as we imagined. Still, it always made the person who wore it look slender. Compared with the loose big-sleeve cheongsam in the 1920s, this kind of cheongsam was slimmer. Due to the improved length and fitting of the cheongsam, the number of buttons on the side increased appropriately. When looking from the side and sitting position, the slit was designed only on the calf, which made a light-colored petticoat with lace hem faintly visible. Such design gave the calf a sense of hazy beauty, as shown in Figure 1-12.

图1-13　阮玲玉穿的高开衩旗袍
Cheongsam with High Slits Worn by Ruan Lingyu

在旗袍长到足面的时代，也有开衩至臀围处的设计，但一般会开至膝盖左右，便于行动。1933年起曾短暂流行过高开衩旗袍，内穿裤子；1934年的《八大女星写真集》中阮玲玉穿着当时最时髦的高开衩旗袍，如图1-13所示；1935年，《妇人画报》中刊登的黄蕙兰穿的假高开衩旗袍，实际开衩至小腿处，如图1-14所示；1937年黄柳霜所穿的高开衩旗袍，内穿长裤，如图1-15所示。

In the era when cheongsam was long enough to reach the feet, there were also styles with slits up to the hip, but usually around the knee, which was convenient for movement. Since 1933, the cheongsam with high slits and pants inside became a fashion for a while. In the *Photo Album for Eight Famous Female Stars* in 1934, Ruan Lingyu, a famous Chinese actress at that time, wore the most fashionable high-slit cheongsam, as shown in Figure 1-13; in 1935, *Women*, a magazine, presented a picture of Huang Huilan who wore a fake high-slit cheongsam that was open to the calf, as shown in Figure 1-14; the high-slit cheongsam worn by Anna May Huang in 1937 was matched with trousers under it, as shown in Figure 1-15.

图1-14　黄蕙兰与假高开衩旗袍
Huang Huilan and the Fake High-Slit Cheongsam

图1-15　黄柳霜穿的高开衩旗袍
High-Slit Cheongsam Worn by Huang Liushuang

2. 旗袍领子/Collar of Cheongsam

20世纪30年代，曳地旗袍的高领设计贴合脖子，起到了支撑作用，增多的扣子也保持了领子的形态。曳地旗袍最为典型的款式便是高领三颗扣，有的款式多至五颗，如图1-16所示。

The high collar of floor-length cheongsam in the 1930s provided support for it to fit the neck, and the increased number of buttons also enabled the collar to maintain a good shape. The most typical representative of the floor-length cheongsam was the high collar style with three buttons, and some had as many as five, as shown in Figure 1-16.

图1-16　20世纪30年代的高领旗袍
High-collar Cheongsam in the 1930s

1937年是变化的转折点，旗袍整体风格变得内敛克制，衣长在缩短，高领在降低，向着20世纪40年代更为舒适且实用的立领高度过渡，如图1-17所示。

The year of 1937 witnessed the turning point for change: the general style of cheongsam became restrained with shortened length, and the collar was designed with lower height, transiting towards more comfortable and practical stand-up collar with lower height in the 1940s, as shown in Figure 1-17.

图1-17　1937年的立领旗袍
Stand-up Collar Cheongsam in 1937

（二）旗袍配饰/Cheongsam Accessories

在领口装饰一枚金属花扣，形状有女花朵、蝴蝶等，使低领旗袍显得更为清丽柔美，明艳大方，如图1-18所示。

Decorating the collar with a metal button in the shape of flowers, butterflies, etc, which made the low-collar cheongsam look more beautiful and elegant, as shown in Figure 1-18.

图1-18　领扣下的装饰
Decorations under the Collar Button

20世纪30年代后，新式进口面料层出不穷，如单色鲜艳的阴丹士林布及提花、印花面料，图案有现代感的花草、条格等，还有轻薄半透明面料裁剪制作的高领长旗袍。旗袍的穿法与搭配更加中西合璧，如与西式外套、大衣、风衣、帽子等搭配，穿玻璃丝袜、高跟皮鞋。发型方面，有烫发或短发，烫发的波浪卷曲充满律动感，而短发则显得率真利索，如图1-19所示。

After the 1930s, new imported fabrics emerged in an endless stream, such as bright monochrome indanthrene cloth and jacquard and printed fabrics, with diverse patterns including modern flowers, stripes, etc. In addition, there was also the high-collar cheongsam cut made of light and translucent fabrics. The wearing methods and matching style succeeded in the integration of the Chinese and Western clothing cultures, such as matching with Western-style coats, overcoats, windbreakers, hats, etc. It can also match with silk stockings and high-heeled leather shoes, with hairstyles varied from perm to short hair that made women more dynamic and forthright, respectively, as shown in Figure 1-19.

图1-19　旗袍配饰
Cheongsam Accessories

二、无袖旗袍/Sleeveless Cheongsam

20世纪40年代初，旗袍的立领高度进一步降低，作为日常化的穿着，旗袍的装饰也趋近于无，如图1-20所示。

In the early 1940s, the height of the cheongsam's stand-up collar was lowered. Meanwhile, as daily wear the decorations of cheongsam, were greatly simplified，as shown in Figure 1-20.

此时，以西式图案为主的各式印花面料制成的无袖旗袍开始流行，常搭配毛衣等外套穿着，如图1-21和图1-22所示。

At that time, sleeveless cheongsams made of

图1-20　20世纪40年代的无袖旗袍
Sleeveless Cheongsam in the 1940s

various printed fabrics with Western patterns became popular. And women often wore it with sweaters and coats, as shown in Figure 1–21 and Figure 1–22.

图1-21 面料图案的西式化
Westernization of Fabric Patterns

图1-22 面料与穿着方式更趋西化
Gradually Westernized Fabrics and Dressing Styles

　　20世纪30年代和40年代的旗袍特点对比如图1-23所示，左边为垫肩双襟绢花装袖旗袍，是30年代初期的样式；中间为无袖紧身前开衩旗袍，是30年代中期的样式；右边为织锦缎无袖双襟旗袍，是30年代末、40年代初的样式。

　　Comparation of the characteristics of cheongsams in the 1930s and 1940s is shown in Figure 1–23, the one on the left is the cheongsam in the early 1930s that has shoulder pads, double lapels, silk flower and sleeves; the one in the middle is the cheongsam in the mid–1930s which is skintight and sleeveless with slits; the one on the right is the style in the late 1930s and early 1940s that is the brocade sleeveless double–lapel cheongsam.

图1-23　20世纪30年代和40年代的旗袍特点对比
Comparison of the Characteristics of Cheongsam in the 1930s and 1940s

20世纪40年代周璇常穿的旗袍据说叫"璇款"，袖子介于无袖和半袖之间，如图1-24所示。

The cheongsam that Zhou Xuan often wore in the 1940s is said to be called "Xuan Style", which fell in between sleeveless and half-sleeved, as shown in Figure 1-24.

图1-24　周璇的无袖旗袍
Zhou Xuan's Sleeveless Cheongsam

第三节　旗袍演变/The Evolution of Cheongsam

一、尖胸旗袍/Pointed-chest Cheongsam

旗袍发展到后来，受内衣和世界时装流行的影响，采用立体裁剪，以尖胸、细腰为突出特点，比如今的旗袍更强调女性的S形曲线，如图1-25所示。且短袖、高领，领口处有明显的圆弧，凸显颈部和面部的秀丽妩媚。收下摆、低开衩，勾勒出外在形体轮廓的错落有致，干脆利落，将人们对女性身体的理想之处明确点出，却无丝毫啰唆累赘之处。旗袍后来的样子依然婀娜而不媚俗，时髦而独立。

With the evolution of cheongsam, it adopted three-dimensional tailoring which featured prominently in pointed chest and thin waist. This style was influenced by the popularity of underwear and global fashion trends, placing more emphasis on the S curve of women than today's cheongsams do, as shown in Figure 1-25. Such style was short-sleeved, high-collar, with a clear arc at the neckline, highlighting the beauty and charm of the neck and face. The closed hem and low slits outlined the delicate body shape, making it look more simplified and beautiful. It gave prominence to the highlights of women's bodies with simple design. Cheongsam in later times remains graceful and noble, fashionable and unique.

图 1-25　尖胸旗袍
The Pointed-chest Cheongsams

尖胸旗袍的领口处通常有明显的 M 形圆弧，在视觉上对脖颈有拉伸效果，如同 V 领、长项链等的作用，使面部秀丽妩媚。

The Pointed-chest Cheongsam usually has an obvious M-shaped arc at the neckline, visually stretching the neck. Additionally, matching the V-neck, long necklace, etc. with the pointed-chest cheongsam can make the face beautiful and charming.

这种高领短袖旗袍刚出现时长度以过膝为主，袖长在腋前寸许，略去胸前的扣子，大襟弧度圆润；不仅立体剪裁，而且使用拉链；下摆依旧低开衩。

When this high-collar short-sleeved cheongsam first appeared, the length was mainly over the knee, the sleeve length was about an inch in front of the armpit, the buttons on the chest were omitted, and the front was rounded; it not only adopted the three-dimensional cutting technique, but also used zippers, with low slits.

尖胸旗袍主要流行于港台地区，直至 20 世纪 60 年代末，旗袍逐渐在中国香港地区淡出日常服饰。中国大陆那时正处于服饰的变革期，仅有少数高层在外交场合穿着旗袍。

The pointed–chest cheongsam was mainly popular in Hong Kong and Taiwan regions. Until the end of the 1960s, a decreasing number of people would choose cheongsam as their daily wear in Hong Kong. At that time, the mainland of China was experiencing a transformation in dress code, and only some high–level executives would wear cheongsams on diplomatic occasions.

日常穿着时，尽管依然保留尖胸旗袍的轮廓，但大大削弱了曲线的强度，领子高度也降低了，胸部曲线也不那么锋利了。面料以素色印花为主，十分简单朴素，如图1-26所示。那时一些欧美国家也印有旗袍款式的纸样，如图1-27所示。

For daily wear, though the form of pointed–chest cheongsam was retained, the curve of the dress was greatly flattened and the height of the collar also lowered, making the curve of the chest not so sharp. The fabric used then was printed in plain colors, as shown in Figure 1–26. At that time, publications with different styles of cheongsams also appeared in some European and American countries, as shown in Figure 1–27.

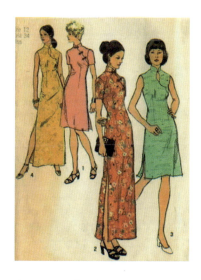

图1-26　素色印花的日常旗袍
Plain Printed Cheongsam

图1-27　国外刊物上的旗袍纸样
Different Styles of Cheongsam in Foreign Publications

二、制服旗袍/Uniform Cheongsam

（一）香港的旗袍式样校服/Cheongsam-style School Uniform in Hong Kong

民国之后，旗袍的款式变化与之前的相比发生了显著变化，如图1-28所示。但"旗袍的校服时代"似乎仍然保持着一股清流。目前，中国香港的许多传统学校仍在使用的旗袍式样校服，大约是在20世纪50年代旗袍的基础上演变而来的。这些旗袍校服采用了无省、装袖、暗扣、侧拉链的设计，如图1-29所示。

After the Republic of China, the style of cheongsam experienced a significant change, as shown in Figure 1–28. But certain characteristics of cheongsam design in "the cheongsam–style school uniform era" seemed to retain to some extent. Currently, students in numerous traditional schools in Hong Kong still wear the cheongsam–style school uniforms, which evolved from the cheongsams in the 1950s. These uniforms are designed with sleeves, hidden buttons and side zippers, without any darts, as shown in Figure 1–29.

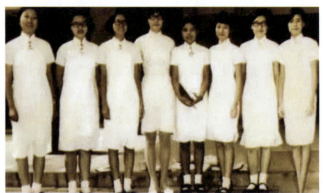

图1-28　20世纪50年代与70年代的香港旗袍式样校服
Hong Kong Cheongsam–style School Uniforms in the 1950s and 1970s

图1-29　现代香港旗袍式样校服
Modern Hong Kong Cheongsam–style School Uniforms

　　与日常服饰不同，制服的变化相对细微和滞后，所以才得以更好地保存了一些细节，尽管制服用途使其显得呆板，但是依然保持着斯文温柔的遗风。旗袍校服一般搭配针织开衫或者套头衫、无袖针织背心等，如图1-30所示。

Different from daily wear, the uniforms have taken on subtle changes, and it took quite a period for such changes to become a common reality. As a result, some details can be better preserved. Despite the dullness in their wearing occasions, these uniforms still pass on the feature of gentleness. Cheongsam–style school uniforms are generally matched with knitted cardigans or pullovers, sleeveless knitted vests, etc., as shown in Figure 1–30.

图1-30　现代香港旗袍式样校服的搭配
The Matching of Modern Hong Kong Cheongsam-style School Uniforms

（二）中国台湾的航空公司的制服设计/Uniform Design of Airlines in Taiwan，China

中国台湾的航空公司曾多次更改空乘制服的设计，但始终不变的是以旗袍作为设计基础，从另一个角度见证了旗袍后来的演变路程。

Airlines in Taiwan，China have changed the design of the flight attendant uniforms several times during its operation, but it stayed true to the basis of cheongsam, which has witnessed the evolution of the cheongsam from another perspective.

20世纪60年代，中国台湾的航空公司制服显现出典雅和性感，如图1-31所示。70年代，中国台湾的航空公司制服在旗袍外面搭配西装式外套，突出职业感，如图1-32所示。70年代也出现了几款斗篷的搭配，有长斗篷也有小斗篷，知性又可爱，如图1-33所示。

The uniforms of Airlines in Taiwan, China in the 1960s were charming in an elegant way, as shown in Figure 1-31. They were matched with a suit jacket on the outside to strengthen the sense of professionalism in the 1970s, as shown in Figure 1-32. Different styles of cloaks also appeared in 1970s, including long cloaks and small cloaks, which were both intellectual and lovely, as shown in Figure 1-33.

图1-31　中国台湾的航空公司20世纪60年代旗袍式空乘制服
The Cheongsam-style Flight Attendant Uniform of Airlines in Taiwan, China in the 1960s

图 1-32　中国台湾航空公司 20 世纪 70 年代旗袍式空乘制服
The Cheongsam-style Flight Attendant Uniform of Airlines in Taiwan, China in the 1970s

图 1-33　中国台湾航空公司 20 世纪 70 年代斗篷与旗袍式空乘制服
The Cheongsam-style Flight Attendant Uniform of Airlines in Taiwan, China with Cloaks in the 1970s

　　20 世纪 90 年代至 21 世纪初，中国台湾的航空公司制服采用蓝紫色调的装饰花扣和立领设计，鲜明利落，至今被很多中式制服借鉴，如图 1-34 所示。

From the 1990s to the early 2000s, the uniforms of Airlines in Taiwan, China designed with blue-purple floral buckles and stand-up collars were distinctive and neat, which have been a great reference for many Chinese uniforms to this day, as shown in Figure 1-34.

图 1-34　中国台湾航空公司 20 世纪 90 年代旗袍元素空乘制服
The Cheongsam-style Flight Attendant Uniform of Airlines in Taiwan, China in the 1990s

思考题

（1）论述旗袍是如何诞生的，你更倾向于旗袍诞生的哪种说法？

（2）简述各时期旗袍的变化特点。

（3）你认为旗袍最有特色的部分在哪里？为什么？

Questions

（1）Talk about how the cheongsam was born, and which statement of its birth do you prefer?

（2）Briefly describe the changing characteristics of cheongsam in each period.

（3）What do you think is the most distinctive part of the cheongsam？ why?

○ 第二章

旗袍造型之思
Thoughts on the Shape of Cheongsam

一、旗袍设计概述/Overview of Cheongsam Style Design

（一）旗袍风格流派/Classification of Cheongsam Styles

旗袍，在不同地域文化的熏染下，也带上了不同的地域色彩。

Cheongsam, influenced by different regional cultures, embodies different regional characteristics.

1. 京派旗袍/Beijing-style Cheongsam

图2-1　京派旗袍
Beijing-style Cheongsam

旗袍本来就是由清代旗女袍服及汉族袍服基础上逐渐发展而来的，所以从保留中华优秀传统文化的角度来看，京派旗袍更加原汁原味地体现了传统文化的精髓，更具有传统旗袍的美感，如图2-1所示。

The cheongsam was originally developed from the banner women's gowns of the Qing Dynasty and the Han robes. Therefore, from the perspective of preserving excellent traditional Chinese culture, the Beijing-style cheongsam is a more authentic reflection of the traditional culture quintessence with the distinctive traditional beauty of such costume, as shown in Figure 2-1.

总体来看，京派旗袍具有重装饰、色彩华丽、绲边很宽等特点。其图案纹饰较为传统，常选择牡丹、梅、兰、青花瓷等图案。面料多以传统绸缎为主，偏厚重，印花面料种类比海派旗袍要少，花色也未受到西方的影响。京派旗袍独有一种端庄传统在里面，给人一种非常成熟的魅力，相较于"海派"的开放与创新，"京派"显得矜持传统一些。

In general, the Beijing-style cheongsam is characterized by heavy decoration, gorgeous colors, and wide piping, with traditional patterns such as peony, plum and orchid, and celadon. Its fabrics mostly consist of traditional satin, which is thick and heavy. There are fewer printed fabrics than Shanghai-style, and the colors have not been influenced by the West. The Beijing-style cheongsam has a unique dignified tradition, showing a kind of mature charm. Compared with the openness and innovation of the "Shanghai-style", this style seems to be more reserved and traditional.

2. 海派旗袍/Shanghai-style Cheongsam

海派旗袍源自上海，在20世纪上半叶由上海女性参考传统服饰和西洋文化设计的一种全新的连身装样式，是东西方文化糅合的体现。所以海派旗袍的最大特点在于对传统服饰与西方服饰的兼收并蓄，如图2-2所示。

The Shanghai-style cheongsam, which originated in Shanghai, was designed by local women in the Republic of China with reference to traditional women's clothing and the Western culture in the first half of the 20th century. In this sense, the most distinctive feature of Shanghai-style cheongsam lies in its eclectic mix of Chinese traditional clothing and Western clothing, as shown in Figure 2-2.

图2-2　海派旗袍
Shanghai-style Cheongsam

海派旗袍设计利用了美学规律和人体工程学原理，并与人体曲线设计相融合。它主要通过突出人体的曲线特征，展现出含蓄、典雅的美感。在设计过程中，采用立体裁剪的方法，在镶边、嵌条、绳边、荡条、盘扣、刺绣等细节上精心制作，从而突出海派旗袍的特色。

The design of Shanghai-style cheongsam conforms to the aesthetic laws and ergonomic principles, which fits well with the human body curve. Such design aims at showing the implicit and elegant aesthetic characteristics by highlighting the human body curve. It adopts three-dimensional cutting, and details, such as trimming, molding, piping, swinging strips, knot buttons, and embroidery，which are meticulously designed to make the style prominent.

3. 苏派旗袍/Suzhou-style Cheongsam

苏派旗袍在吴文化的长期浸润下，工艺中大量融入了精美的苏绣、缂丝等苏州代表性的非物质文化遗产技艺；在选材方面，除了常见的丝绸面料外，还倾向于使用苏州地区生产的传统面料，如宋锦、苏罗等。此外，苏派旗袍与绘画艺术紧密相连，作为"吴门画派"的发

图2-3　苏派旗袍
Suzhou-style Cheongsam

源地，苏州的绘画艺术繁荣昌盛，这使苏州地区的旗袍能够大量采用刺绣、手绘等方式来展现图案，如图2-3所示。

Under the long-term influence of Wu culture, Suzhou-style cheongsam has incorporated a great many techniques in the exquisite Su embroidery, Kossu (a type of weaving done by the tapestry method in fine silks and gold thread) and other representative intangible cultural heritage techniques of Suzhou. In terms of material selection, apart from the common silk fabrics, the traditional fabrics produced in Suzhou are preferred, such as Song Brocade, Su Luo, etc. In addition, it is intricately associated with painting. Suzhou, as the birthplace of the "Wumen School of Painting", boasts a long history in painting, which offers inspiration to the design of Suzhou-style cheongsam with embroidery, hand-painting and other methods to display the patterns, as shown in Figure 2-3.

苏州地区的女性旗袍风格源于精致考究、娴雅舒适的苏式生活，它映衬了苏州女性优雅、温婉的品格，并带有独特的苏州符号与浓郁的"苏州味道"。这种服装风格自然与"京派"的古典、大气和"海派"的摩登、浪漫形成鲜明对比。

The style of women's cheongsam in Suzhou area originated from the exquisite, elegant and comfortable Suzhou lifestyle, reflecting the graceful and gentle character of Suzhou women, with unique Suzhou symbols and regional characteristics of Suzhou. Its clothing style varies naturally from the traditional, classical Beijing style and the modern, romantic Shanghai style.

（二）旗袍设计元素/Cheongsam Design Elements

1. 旗袍的"点"/The "Point" of Cheongsam

在世人眼中，旗袍是最能体现东方古典美的服饰，它最贴合东方女性的性格。旗袍的外部特征中，最能体现其精致之处的便是"点"——盘扣与立领，这两个巧妙精致的设计元素，是旗袍外在形式美的点睛之笔。盘扣，又称盘钮、布扣，是由古老的"结"发展而来。在古代，严苛的封建礼教束缚着中国女性，当时妇女的服饰剪裁线条都以平直为主，为避免暴露身形曲线，衣服多肥大宽平。因此，在旗装这一平面服饰上，盘扣就成为除面料纹样以外唯一的设计空间。那时的妇女喜欢在领口、袖口、披襟上加上颜色艳丽、形状精致的盘扣来装点衣袍。从栩栩如生的蝴蝶扣、蜻蜓扣、菊花扣到象征吉祥的寿形扣、云纹扣、如意扣，盘扣的装饰意义与象征意义已经远远超出了其实用功能，而由于其经典的外部造型特征，经常

被用于"中国风"的视觉设计作品中，如图2-4所示。

In people's eyes, cheongsam is the clothing that best reflects the classical beauty of the East, best fits the character of oriental women, and it is also the most exquisite costume. Among the external features of the cheongsam, the most exquisite design lies in the Chinese knot button and the stand-up collar. These two ingenious and delicate designs are the highlight of the external form of cheongsam. The Chinese knot button（Pan Kou）, also known as the Pan Niu and Bu Niu, was developed from the ancient "knot". Chinese women in ancient times were restricted by austere feudal ethics, and their clothing was loose and flat, cut with straight lines to avoid the exposure of body curves. In such context, the knot button has become the only spot to showcase delicate designs other than the fabric pattern for the banner gowns. At that time, women loved to decorate their garments with colorful and delicate knot buttons on the collar, cuffs, and front of the clothing. These knot buttons, including butterfly buttons, dragonfly buttons, chrysanthemum buttons, and those symbolizing good luck, like the Shou character-shaped buttons（the Chinese character "寿" symbolizes longevity）, cloud-shaped buttons, Ruyi buttons, etc., are used more for decorative and symbolic purpose than their original function. Its classical external features make the knot button a common element in "Chinese style" visual design works, as shown in Figure 2-4.

 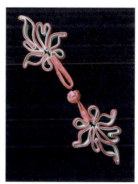 

图2-4　盘扣的视觉作用
The Visual Effect of the Knot Button

旗袍的立领是其显著的标识。在旗袍的改良过程中，领子也经历了多种样式的演变，最终发展为固定的领型，如一般领、企鹅领、凤仙领、水滴领、竹叶领、马蹄领等。作为旗袍的标志性符号，立领元素也是常在设计作品中出现的一个外形构造，如图2-5所示。

The stand-up collar of cheongsam is a very conspicuous symbol on the costume. During its improvement process, the collar has also undergone various evolutions, and eventually，it has now developed to a fixed style. There are different shapes of the stand-up collar, for example：the common collar, the penguin collar, the balsamine-shaped collar, the water-drop collar, the bamboo-leave collar, and the horse-hoof-shaped collar. As the iconic symbol of cheongsam, the stand-up collar element is also a shape that often appears in design works, as shown in Figure 2-5.

图2-5　领子的视觉作用
The Visual Effect of the Collar

2. 旗袍的"线"/The "Line" of Cheongsam

旗袍的经典标志除了立领与盘扣，还有一个显著特征便是旗袍玲珑妩媚的曲线。在旗袍的视觉整体中，灵动的线条串联起各部分构造，使旗袍的外在特征在视觉上达到平衡。旗袍的"线"包括旗袍的剪裁轮廓、裙摆开衩、大襟三个部分，它们如同画面上的线条，勾勒出旗袍的形态，平衡整体画面。旗袍的剪裁轮廓是一个由"平面"向"立体"转变的过程。传统旗袍的剪裁轮廓是平面的几何T字形，主要采用平直的线条，且衣片与袖片取自同一块布料。如图2-6所示，属于平面剪裁，也就是连身袖结构，这种袖子与衣身连为一体的剪裁方式使旗袍的外部造型概括感极强，但线条略显生硬，缺乏灵动感。

In addition to the stand-up collar and the Chinese knot button, the exquisite and charming curve is another significant feature of cheongsam. The flexible lines get every part of the clothing connected to give a perfect visual effect on the whole, enabling the external features to strike a balance visually. The "line" of the cheongsam includes three parts: the silhouette of the cheongsam, the slits on the hem, and the big front. These three parts are like lines on the picture that delineate the costume's form and balance the overall picture. The silhouette cutting is a process of transformation from "plane" to "three-dimensional". The silhouette of traditional cheongsam was a flat geometric T-shape, mainly adopting straight lines, and the garment and sleeve were made of the same fabric. Figure 2-6 displays flat cutting, that is, one-piece sleeves structure. Such a cutting method well connected the sleeves with the main body in one piece, making the external silhouette look complete, but the lines are too rigid without proper flexibility.

图2-6　传统旗袍的T字形平面裁剪
T-shaped Flat Cutting of Traditional
Cheongsam

民国初期，受外来风尚影响，旗袍的剪裁轮廓也发生了改变，中式平面剪裁开始向西式立体剪裁慢慢转变，这种"立体"的剪裁方式是一种模拟人体穿着状态的分割剪裁方式，使衣料更加贴近人体曲线，自此，旗袍的剪裁轮廓变为模仿女性优美曲线的"S"形，这种"S"形的轮廓线也成了近代旗袍的一个重要标志。

In the early days of the Republic of China, the cutting pattern of cheongsam also changed under the influence of foreign fashion trends. The Chinese flat cutting method began to take on the Western−style three−dimensional cutting – a separating cutting method that imitates the wearing status of the human body, making the fabric more close−fitting to the curve of the human body. Since then, the silhouette of cheongsam has become an "S" shape that imitates the graceful curves of women, which also marks an important feature of modern cheongsam.

3. 旗袍的"面"/The "Appearance" of Cheongsam

旗袍的"面"涵盖了面料、纹样、配色三个部分。这是在旗袍整个外部形态中所占面积最大的一部分，结合前文所说的"点"与"线"两个部分构成旗袍的立体视觉外观。旗袍的面料主要有织锦、棉布、丝绒、真丝等。面料的选用奠定了旗袍整个外部形象的基调，直接影响其整体质感与档次。

The "appearance" of cheongsam mainly contains three parts: the fabric, the pattern, and the color. The appearance constitutes the largest part of the entire external shape of cheongsam, which, coupled with the "point" and "line" mentioned above, form the three−dimensional visual appearance of the clothing. Cheongsam's fabrics include brocade, cotton, velvet, silk and so on. The selection of fabric sets the tone for the external image of cheongsam, which affects the texture and grade of the final product.

旗袍的纹样种类繁多，不仅造型具有民族特色，还代表了中国民众从古至今的"吉祥观"。虽然不同时期旗袍的纹样整体风格各不相同，但是所选用题材大致相同，多围绕寓意吉祥富贵的事物，寄托着中国人民的美好愿望与中国女性的女儿情思。中国古代追求"天人合一"，喜欢赋予万物以"意"，让自然万物代表人民的情感。旗袍的纹样也同样汲取自然之"意"，纹样上的山水花鸟，便是"意"的衍生，借天地万物歌颂君子淑女之德。

Cheongsam features a wide variety of patterns, which not only boast national characteristics, but also represent the "auspicious concept" of the Chinese people from ancient times to the present. Although the cheongsam patterns vary in different periods, the selection of cheongsam patterns is roughly the same, focusing on things that imply auspiciousness and wealth, expressing the aspirations of Chinese people, and the affection of Chinese women. Ancient China emphasized "the unity of man and nature", and liked to give all things "meaning" and let all things in nature represent people's feelings. The cheongsam patterns also capture the inherent "meaning" of nature in life. The landscapes, flowers and birds on the patterns are the derivations of "meaning".

除了中国古代文人思想"意"的影响，古代民俗文化也对旗袍纹样的题材选择有着重要影响。旗袍是生活用品，自然人们在生活层面的风俗习惯也会渗透在里面。纹样的选择既涉及创意造物，又反映生活的方方面面，如祭祀、庆典、婚丧嫁娶……这些纹样既装饰了旗袍的脸面，又体现了使用者的意念与情思。比如鸳鸯或并蒂莲寓意爱情与夫妻和美，葡萄和石榴寓意多子多福，牡丹凤凰寓意吉祥富贵等。这些寓意丰富的旗袍纹样为当代设计提供了丰富的资源与灵感。旗袍纹样不仅使旗袍的外部形态有了鲜明的风格特点，还使旗袍的整体外部视觉形态丰富充盈起来，如图2-7所示。

In addition to the influence of the "meaning" of ancient Chinese literati thought, ancient folk culture also has an important influence on the choice of the theme of the cheongsam patterns. Cheongsam is a type of daily necessity, so naturally the customs and habits of people in life will be iconic reflections of cheongsam. The selection of patterns involves all aspects of life, such as sacrifice, celebration, wedding, funeral, etc. These patterns not only decorate the appearance of cheongsam, but also express the aspirations and feelings of the wearer. For example, the mandarin duck or the lotus symbolizes love and the harmony between husband and wife, the grapes and pomegranates symbolize more offspring and blessings, and the peony and phoenix symbolize auspiciousness and wealth. These cheongsam patterns with rich meanings provide a wealth of resources and inspiration for contemporary design. The pattern of cheongsam not only gives the appearance of cheongsam unique characteristics, but also enriches the visual form of cheongsam, as shown in Figure 2-7.

图2-7　传统旗袍的"面"
The "Appearance" of Traditional Cheongsam

二、制服类旗袍的设计/Design of Uniform Cheongsam

（一）制服类旗袍的设计要点/Design Essentials of Uniform Cheongsam

（1）制服类旗袍设计的四要素即职业性、经济性、审美性、功能性。

Four design elements of uniform cheongsam: occupational, economical, aesthetic and functional.

（2）制服类旗袍设计首先要有明确的针对性，这包括对不同行业、同一行业不同企业、同一企业不同岗位、同一岗位不同身份和性别等的考量。同时在同等的美感与功能前提下，设计工作服要尽可能降低成本，从款式、材料、制作的难易程度、服装的结构等细处着眼。只有做到四要素俱全，才能达到整体设计的最佳效果。

The design of uniform cheongsam must first be targeted at different industries, including different companies in the same industry, different positions in the same company, different identities, genders and so on in the same position. Meanwhile, under the premise of the same aesthetic experience and function, the design of uniforms should reduce the cost as much as possible, focusing on the details of style, material, difficulty of production, and structure of clothing. Only when the four elements are met to the largest degree can the design achieve the overall effect.

（3）制服类旗袍创新三要求：设计风格创新、款式创新、搭配创新。

Three requirements for uniform cheongsam innovation：design innovation, style innovation, and matching innovation.

（二）制服类旗袍的创新设计/Innovative Design of Uniform Cheongsam

1. 廓型的创新设计/Innovative Design of the Silhouette

旗袍的廓型设计是影响其整体变化和视觉印象的关键因素。现代旗袍除 A、H、X 这三种基本廓型外，经过设计师的创新设计变得更加多元化，O 形、T 形、V 形，甚至异形轮廓都被应用到现代旗袍的外轮廓设计中。在进行廓型设计时，要从二维平面和三维空间的关系出发，协调好整体的比例关系，运用意象、联想、移用、夸张、转移、变换等多种创新设计方法进行整体设计。移用法是指通过对已有设计元素进行有选择地吸收、融合，从而形成新的设计方案，有直接移用法和间接移用法。2008 年北京奥运会颁奖礼仪服装中就采用直接移用法，将中国的青花瓷造型和图案运用到礼服旗袍的设计中，如图 2-8 所示。

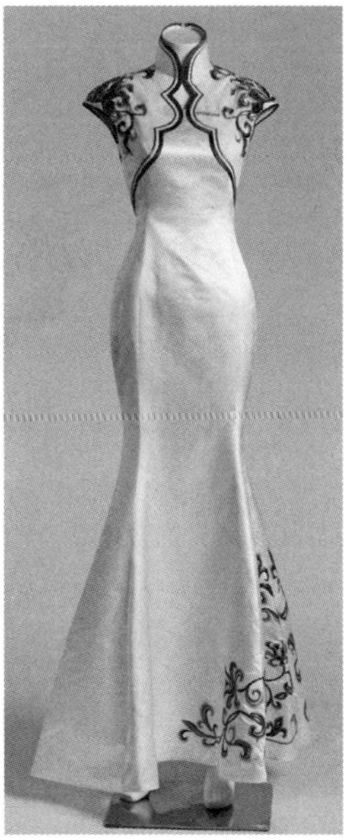

The silhouette design of cheongsam is a key factor affecting the overall change and visual impression of the clothing. In addition to the three basic silhouettes of A, H, and X, modern cheongsam has become more diversified due to designer's innovation. O-shaped, T-shaped, V-shaped, and even special-shaped contours have been applied to the contour design of modern cheongsam. Such a design shall start from the relationship between the two-dimensional plane and the three-dimensional space, coordinate the overall proportion, and use a variety of innovative design approaches such as imagery, association, transferring, exaggeration, and transformation to serve the overall design. The transferring method refers to the selective absorption and fusion of existing design elements to form a new design scheme, including direct transferring and indirect transferring. For instance, the award ceremony clothing in the 2008 Beijing Olympic Games took the

图 2-8　青花瓷旗袍造型
Blue and White Porcelain
Cheongsam

direct transferring method, applying the Chinese blue and white porcelain shapes and patterns to the design of cheongsam, as shown in Figure 2–8.

2. 局部造型的创新设计 /Innovative Design of Parts

（1）衣领的创新设计。立领是一种将领片竖立在领围线上的领型，又称竖领。传统旗袍的领型一般有立领、上海领、凤仙领等，现代旗袍设计在保留立领基本造型的基础上，运用夸张、变形、分割、镂空等造型手法，对旗袍领型的三维结构或某一局部进行创新设计。如可采用镂空的造型手法，对镂空的形状、材料、表现形式等进行设计。例如，旗袍设计运用了借鉴法，将古代窗格元素运用到旗袍设计上，赋予其浓郁的中国情调。在领部设计中，将窗格元素打散重构，利用其图案的秩序感向下延伸，勾勒出镂空的胸、腹部造型。窗格元素的应用形成了重复与对称的视觉感受，面料的冲突感增加了整体的韵律感，使设计在形式上充满节奏感与韵律感，同时将传统元素中所蕴含的深厚的古代文化传达给了观者，如图2-9所示。另一种创新设计是立领无肩造型，立领的方正严密与肩部的裸露形成鲜明对比，增强了旗袍的现代感。

Innovative design of the collar. A stand–up collar is a type of collar in which the collar piece is erected on the collar line. The collar types of traditional cheongsam generally include stand–up collar, Shanghai collar, balsamine–shaped collar, etc. The modern cheongsam design still retains the basic shape of stand–up collar, based on which exaggeration, deformation, division, hollowing and other methods are used to innovatively design the collar or certain parts. For example, the cheongsam design refers to the ancient pane to showcase Chinese culture. Dismantling and reconstructing the pane elements, and stretching downwards by taking advantage of the pattern's

图2-9　旗袍局部设计——领子
Partial Design of Cheongsam—Collar

order, the hollow chest and abdomen shape can be outlined. The application of pane elements renders a visual effect of repetition and symmetry. The conflict of fabrics increases the extension of rhythm in form, and conveys the profound ancient culture contained in traditional elements in the meantime, as shown in Figure 2–9. It can also be designed as a stand–up collar with no sleeves. The square of the stand–up collar is closely contrasted with the exposed shoulders, enhancing the modernity of cheongsam.

（2）开襟的创新设计。从商、周时期开始，中国的传统袍服便采用了开襟的形式，依据各民族的生活习惯和生活方式的不同将门襟分为右衽门襟、左衽门襟和对襟三种。中国自古以来就有"以右为尊"的传统，故服饰以"右开襟"为主。旗袍的开襟形式主要有斜襟、曲襟、双襟、方襟、如意襟、琵琶襟等。这些开襟形式在服装造型中起到分割线作用的同时，还起到装饰美化的作用。

Innovative design of the placket. Since the Shang and Zhou Dynasties, traditional Chinese gowns adopted the design of lapel（open side of the clothing）which was divided into three types：right lapel, left lapel and double lapel, in accordance with the different living habits and lifestyles of various ethnic groups. Chinese people ever since ancient times have always placed what they respect on the right side, therefore, the "right lapel" became dominant. There are different styles of lapel, such as the slanting lapel curved lapel, double lapel, square lapel, Ruyi lapel, Pipa lapel, etc. It is not only a dividing line of the clothing, but also a decoration of it.

　　现代旗袍的门襟设计形式多样，已打破了传统右开襟的束缚，设计变得更加灵活大胆。襟部的设计主要从门襟曲度、襟位变化和襟缘装饰入手。襟形可设计为不规则形状，门襟的位置可以灵活移动，也可通过使用异质面料来增添门襟的趣味感。另外，对盘扣的创新设计更是起到了画龙点睛的作用。现代旗袍的设计追求和而不同，同中求异，如图 2-10 所示。

　　The lapel design of modern cheongsam has various forms, breaking the traditional form of the right Lapel, and the design has become more flexible and bold. Its design focuses on its curvature, the change of its position and the decoration. The Lapel can be designed in an irregular shape, with a movable position and made of exotic fabrics, which makes it interesting. Furthermore, the innovative design of the knot button is a highlight. The design of modern cheongsam emphasizes the co-existence of harmony and difference, as shown in Figure 2-10.

图 2-10　旗袍局部设计——门襟
Partial Design of Cheongsam—Lapel

　　（3）下摆的创新设计。现代旗袍裙摆造型多借鉴礼服的裙摆样式，常见的下摆设计有鱼尾式、燕尾式、曳地式等。裙摆设计是现代旗袍设计的一大亮点，这些变化形式多通过对裙部基本型的分割、移位、展开与变形来实现。例如，在腰节处增加横向分割线断开旗袍的连身结构，通过立裁的方式改变裙部造型；也可运用夸张的设计手法加大、加长裙摆，从而强调整体的体积感与层次感；还可运用重复、叠加的设计手法使裙摆产生节奏感，如图 2-11 所示。

　　Innovative design at the hem. The hemline style of modern cheongsam mostly refers to that of the ceremonial dress, varying from fishtail style, swallowtail style, floor dragging style, etc. The

图 2-11　旗袍局部设计——下摆
Partial Design of Cheongsam—Hem

hemline design is a highlight of the modern cheongsam, changing its shapes by dividing, shifting, unfolding and deforming the basic shape of the hem. For example, a horizontal dividing line is added at the waist to divide the one-piece structure, and the shape of the hem can be changed by vertical cutting; exaggerated methods can also be used to enlarge and lengthen the hemline to strengthen the sense of layering; The effect of repetition and overlays can be used to create a sense of rhythm in the hemline, as shown in Figure 2-11.

（三）制服类旗袍的设计案例/Design Cases of Uniform Cheongsam

1. 北京奥运会礼仪服装设计/Beijing Olympics Etiquette Clothing Design

2008年北京奥运会颁奖礼仪服装共推出了男装1款，女装5个系列共15款。前者刚柔相济，后者柔美典雅，它们共同亮相于北京奥运会和残奥会的共七百余场颁奖仪式上。

At the 2008 Beijing Olympics, a total of 15 styles for men and five series for women were rolled out as awards ceremony clothing. The former is a perfect combination of rigidity and softness, while the latter is soft and elegant, debuting at more than 700 award ceremonies in the Beijing Olympic and Paralympic Games.

这5个系列女装分别为"青花瓷"系列、宝蓝系列、国槐绿系列、"玉脂白"系列和粉红色系列。每一系列都根据嘉宾引导员、运动员引导员和托盘员的不同职能，精心设计了三个不同款式。国际奥委会对北京奥运会颁奖礼仪服装给予了高度评价，认为它们展现了中国特色。

The five series of women's clothing consist of the "Blue and White Porcelain" series, the sapphire blue series, the Chinese scholar-tree green series, the jade-white series and the pink series. Each series has three different styles according to the different duties of the wearers. The International Olympic Committee spoke highly of the Beijing Olympics award ceremony clothing, regarding it as a full expression of Chinese culture.

"青花瓷"系列的设计灵感源自享誉世界的中国青花瓷器。中国传统乱针绣的运用形象逼真地再现了青花瓷的晕染效果，鱼尾裙的廓型设计完美地凸显了中国女性的柔美曲线，如图2-12所示。

The design of the "Blue and White Porcelain" series was inspired by the world-famous Chinese blue and white porcelain. Such patterns are vividly presented in the cheongsam owing to the traditional Chinese embroidery with overlapping threads in different directions. The silhouette design of the fishtail skirt highlighted the soft curves of Chinese women, as shown in Figure 2-12.

宝蓝系列采用温润典雅的宝蓝色作为礼服主色调，腰间饰有采用传统盘金绣制作的精美腰封。腰封上的图案选用了最具中国优秀传统文化审美意趣和美好愿望的吉祥纹样——江山海牙纹、牡丹花纹，不仅凸显了鲜明的中国特色和民族风格，同时也向全世界展示了中国精湛的刺绣技艺。中式的立领配以西式的肩部设计尽显中国女性落落大方的高贵气质。在体操、室内球类比赛和击剑等项目的颁奖现场，身着宝蓝色礼服的颁奖仪式专业志愿者成为一道亮丽的风景线，如图2-13所示。

The sapphire blue series, with the warm and elegant sapphire blue as the main color, and a decorated waist covering made by traditional gold embroidery stood out the distinctive Chinese characteristics and national style. It also showed people around the world the most superb embroidery craftsmanship in China. The Chinese-style stand-up collar, coupled with the Western-style shoulder design, unfolded the elegant and noble temperament of Chinese women. Professional volunteers in royal blue dresses appeared at the awards ceremony for gymnastics, indoor ball games and fencing. As shown in Figure 2-13.

图2-12　奥运会系列设
计——青花瓷
Olympic Series Design—
Blue and White Porcelain

图2-13　奥运会系列设
计——宝蓝系列
Olympic Series Design—
Sapphire Blue Series

国槐绿系列以丝缎为面料，寓意蓬勃朝气的生命和郁郁葱葱的环境，体现了人类与自然和谐发展的美好愿望及坚守"绿色奥运"的决心。立体银线绣制的吉祥牡丹和卷曲花纹，与女性柔美曲线相契合，更显身段的婀娜多姿和东方女性的恬静气质。在自行车、射击、现代五项等项目的颁奖仪式上，国槐绿系列成为颁奖礼仪人员的首选，如图2-14所示。

The green (the same color as the Chinese scholar tree) series of dresses made of silk satin symbolize vigorous life and lush environment, reflecting the good wishes of harmonious development between human and nature, and the determination to adhere to the "green Olympics". The auspicious peony embroidered with three-dimensional silver thread and the curly pattern that fits the feminine curves of women show the graceful figure and the tranquil temperament of oriental women. This series was used in award ceremonies for cycling, shooting, modern pentathlon, etc., as shown in Figure 2-14.

"玉脂白"系列巧妙地呼应了奥运奖牌金镶玉的设计理念，彩绣腰封和玉佩的加入，既是对中国玉文化的完美再现，又是对传统旗袍设计的一次创新。层次丰富的绿色与牙白色丝绸面料的质感完美搭配，更突出了中国女性内敛、含蓄的特质。该系列礼服在国家体育场、所有的室外球类比赛及香港马术比赛中亮相，赢得了广泛赞誉，如图2-15所示。

The white (the same color as the mutton fat jade) series perfectly echoes the design concept of Olympic medals that were gold inlaid with jade. The design of a colorful embroidered girdle and jade pendant was not only a perfect reproduction of Chinese jade culture, but also an innovation of traditional cheongsam. The perfect match of green color and the texture of tooth-white silk fabrics highlighted the restrained and reserved characteristics of Chinese women. The series appeared in the National Stadium, all outdoor ball games and the Hong Kong Equestrian Games, as shown in Figure 2-15.

图2-14　奥运会系列设计——国槐绿系列
Olympic Series Design—Green Series

图2-15　奥运会系列设计——玉脂白系列
Olympic Series Design—White Series

图2-16　奥运会系列设计——粉红色系列
Olympic Series Design—Pink Series

粉红色系列以传统盘金绣工艺制作的宝相花图案作为腰饰，分割出完美的人体比例。领部设计凸显出颈部的优美线条。粉色系列服装主要出现在拳击、举重、摔跤等力量型比赛的颁奖场合，以其柔美的粉色调来中和比赛的阳刚之气，如图2-16所示。

In the pink series, the waist ornament with the pattern of composite flowers made by the traditional gold embroidery process divided the human body into perfect proportion. The design of the collar highlighted the graceful lines of the neck. The pink series of clothing was used in the awarding ceremony for games like boxing, weightlifting, wrestling, etc. The color can neutralize the masculinity of the competition, as shown in Figure 2–16.

2. G20国宴礼仪服装设计/G20 State Banquet Etiquette Costume Design

旗袍作为中国的国粹，在国宴上的礼仪服款中占据重要地位。在2016年杭州G20国宴上，设计师为礼仪小姐选择了大方、素雅的旗袍作为制服。这款旗袍的设计关键词是"端庄大气"和"大家闺秀"，旨在展现主人翁的姿态。

As the quintessence of China, cheongsam must be the first choice for the ceremonial dress at the state banquet. So was the design of the uniform for the Miss Manners of the 2016 Hangzhou G20 State Banquet Etiquette, which was elegant and dignified. Such characteristics constituted the key–words of this design, "Duan zhuang Da Qi（dignified and elegant）" and "Da Jia Gui Xiu（lady from a wealthy family）", that showed the temperament of China as the master.

制服以杭州丝绸和最具代表性的西湖风景来展开设计，整体造型摒弃繁复，以简洁为主基调。设计师对款式进行了改良——上半身为旗袍，下半身为裙子，这种两件式的设计既能修饰身材，又便于礼仪场合的行走，如图2-17所示。

The uniform, designed with Hangzhou silk and the most representative West Lake scenery, abandoned complex decoration and focused on simplicity on the whole. The improved style by the designer, with two pieces of upper cheongsam–style shirt and skirt in the lower part not only made the figure more beautiful, but also allowed the wearers to walk more conveniently and flexibly, as shown in Figure 2–17.

图2-17　国宴礼服设计
State Banquet Etiquette Costume Design

3. 广州亚运会礼仪服装设计/Guangzhou Asian Games Etiquette Clothing Design

2010年广州亚运会举牌礼仪服装和运动会颁奖礼仪服装设计同样以旗袍为基本造型，并在此基础上进行了演化创新。其纹样充满了云、水、龙、凤等中国元素，色彩搭配丰富多样，如图2-18所示。

The design for ceremonial clothing and sports awarding ceremony clothing in the 2010 Guangzhou Asian Games took the cheongsam style as the basic shape for further evolution and innovation. Chinese elements such as clouds, water, dragons and phoenixes can be seen everywhere in such design with a rich mix of colors, as shown in Figure 2-18.

图2-18 广州亚运会礼服
设计
Guangzhou Asian Games Etiquette Clothing Design

4. 海南航空公司空乘制服设计/Hainan Airlines Flight Attendant Uniform Design

2020年海南航空发布的第五代机长、乘务员制服以旗袍为原型，在印花图案上，采用中国传统元素，领口为祥云漫天，下摆为江崖海水，寓意翱翔于云海之间，构成独具海南航空特色的东方之美。整体来看，第五代新制服既保持了经典的中国元素，又融入了当下的国际时尚元素，兼具现代美感，如图2-19所示。

In 2020, Hainan Airlines rolled out its fifth-generation captain and flight attendant uniforms. The new uniforms, based on the design of cheongsam, adopted traditional Chinese elements on the printing pattern, with clouds in the collar and rivers in the hem to symbolize blessings. It constitutes the beauty of the East with the unique characteristics of Hainan Airlines. On the whole, this uniform maintains the classic Chinese elements, and integrates the current international fashion elements, perfectly showing the modern beauty, as shown in Figure 2-19.

图2-19 海南航空公司空乘制服设计
Hainan Airlines Flight Attendant Uniform Design

三、生活类旗袍的设计/Design of Daily-wear Cheongsam

（一）旗袍文化内涵在生活类旗袍上的转化/The Transformation of Cheongsam Cultural Connotation in Daily-wear Cheongsam

旗袍的内敛之美在于裹住女性丰腴的胸部、纤细的柳腰和饱满的臀部，从小巧的领口到衣摆顺直长掩到足面，不露声色地显露出女性修长的身姿，它所展现的婉约韵味，是其他服饰所无法比拟的；旗袍的张扬则体现在或高或低的衣袖、起起伏伏的裙边，不经意间袒露出手臂和小腿，既不失优雅，又含蓄内敛。

The cheongsam represents a kind of clothing culture that's implicit and restrained. It wraps around the plump breasts, slender waist and full buttocks of women, and goes straight to cover the feet from the small neckline to the hem, underlining their slender figures in an undemonstrative way. The graceful charm of cheongsam is unmatched by other costumes. On the other hand, the cheongsam is flamboyant, with high or low sleeves and undulating hems, inadvertently revealing the arms and calves of women to show their grace.

生活类旗袍，作为中国千年服饰文化的经典之作，已然成为中华民族的一个标志。它历经岁月的洗礼及人文的变迁，已成为中华民族宝贵的服饰文化精神财富和文化财富。它的整体造型风格体现了中国服饰艺术文化的特质，是中华民族服饰文化的瑰宝。生活类旗袍上局部与整体造型的统一展现了中国人所追求的和平之美；被遮住的颈和胸，与中国人含蓄、内敛的性格相得益彰。同时，现代旗袍设计把中国传统服装文化所蕴含的"意蕴"与西方审美文化所追求的"形"相结合，呈现出中西方文化交融的独特魅力。

As a classic of China's long-standing clothing culture, the daily-wear cheongsam, regarded as a symbol of Chinese nation, has become the precious spiritual and cultural wealth Chinese nation's clothing culture after undergoing a profound humanistic transformation in its long history. Its overall style conforms to the characteristics and is the crystallization of Chinese costume art and culture. The unity of partial and overall styles of such cheongsam shows the peaceful beauty of the Chinese people. Additionally, the covered neck and chest are also consistent with the Chinese people's reserved and restrained characters. And the integration of the implicit traditional Chinese clothing culture that emphasizes on "connotation" with the Western aesthetic culture that pursues the beauty of "shape" contributes to the birth of the modern cheongsam.

传统的旗袍是用观念去穿的，是用一块"精神的布"把身体遮蔽起来，无论造型、色彩还是纹饰，都有其象征意义。传统旗袍以款式宽博、装饰繁冗为美，着装讲求精致，既不过于暗淡，又不过于张扬，以"儒雅"为美。现代生活类旗袍则不同于传统旗袍，由于结合了当代服装发展的潮流，在西式时装元素的影响下，现代生活类旗袍在形式上更强调自然、简洁，不过于雕琢、堆砌。在思想上，现代生活类旗袍追求表达自我的形体美，使女性更显优雅、端庄。

The traditional cheongsam places high value on spiritual culture, aiming at covering the body with a piece of "spiritual cloth". Therefore, everything of such cheongsam, regardless of shape, color, or decoration, has its symbolic meaning. This kind of cheongsam is beautiful with its loose style and complicated decoration, preferring exquisite style that is neither too dim nor too flamboyant, thus delivering the beauty concept of grace. Different from this, the modern daily-wear cheongsam, as a result of the contemporary clothing development and influenced by Western fashion, emphasizes more on the naturalness and simplicity in form with appropriate decoration to liberate women's minds. What the modern cheongsam pursues is the physical beauty of self-expression rather than the complex design, intending to make women more elegant and dignified.

在当今审美多元化的背景下，我们应以现代人的审美眼光去解读生活类旗袍所蕴含的文化内涵，让不完美的形体在人的创作下完美化，从而使其成为全球共享的精神及文化财富。

In today's diverse aesthetic background, developing and interpreting the cultural connotation of the daily-wear cheongsam with the modern aesthetic perspective can make the imperfect body perfect under the creation of people, so as to make it a spirit and cultural wealth shared by the world.

中国旗袍经历了百年的演进发展，经历了沉寂的低沉，更焕然一新地再现。在不同的时代，展现出了各自的个性时尚与服饰文化，旗袍更是传统和现代的纽带，将过去与现在紧紧联结在一起，密不可分。生活类旗袍的款式、面料、色彩都丰富多彩，随着复古风的再度盛行，它将在世界舞台重新展现昔日的风采。旗袍作为中国服饰文化的重要组成部分，不再仅仅是一件服饰，更是一种文化和精神的象征，一种潜在的魅力，一种无以言表的骄傲，如图2-20所示。

图2-20 生活中的旗袍风采
The Style of Cheongsam

In its one-hundred-year evolution and development, the Chinese cheongsam has experienced downs but welcomes a vigorous rebirth now. It not only shows the individual fashion and clothing culture of each era at different times, but also links the tradition and modernity, the past and the present. The styles, fabrics and colors of daily-wear cheongsam are rich and colorful. With the re-emergence of retro style, it will re-interpret its splendor in former times on the world stage. As an important part of Chinese clothing culture, cheongsam is no longer a simple costume, but a symbol of culture and spirit, a potential charm, and great pride, as shown in Figure 2-20.

（二）生活类旗袍的魅力之美/The Beauty and the Charm of Daily-wear Cheongsam

生活类旗袍在长期演化中形成了自己独特的形式美法则。它将东方女性的优雅、柔美、温润、娴雅、妩媚的性情和气质刻画得淋漓尽致、尽显无遗。旗袍与女性的形体和气质相互融合，有着神奇的魅力，完美地把女性的优雅展现出来。穿上旗袍，身体的各个部位形成众多曲线，这些曲线巧妙地结合成一个整体，共同塑造女性独特的魅力。这种自然的廓型与国际上强调的夸张线条截然相反，平和而不失典雅，这便是其魅力所在。生活类旗袍的魅力还在于其不断吸收外来时尚元素，使其独特的神韵与现代审美观念相契合，成为潮流的中心。

Daily-wear cheongsam has established its own unique formal beauty that's spiritual in the long-term evolution. It depicts the elegant, soft, gentle, and charming temperament of oriental women incisively and vividly. Owing to its good integration with women's bodies and temperament, it shows the elegance of women in a perfect way. Putting on the cheongsam, various parts of the body form numerous curves, which are subtly combined into a whole to reflect the charm of women. Diametrically opposed to the exaggerated lines that emphasize curves in the world, the natural silhouette of this costume is elegant in peace and can better demonstrate women's charm. What makes the daily-wear cheongsam so attractive lies in its continuous absorption of foreign fashion elements, making its unique charm and modern aesthetic concepts in common, and becoming the center of the trend.

1. 曲线之美/The Beauty of Curves

旗袍的美是独一无二的，充满了古典的灵性，它与着装者的形体和气质相互融合、映衬，把女性温柔、婉约、性感完美地融合在一起，具有东方式的含蓄风格。旗袍能够自然而然地勾勒出女性的身体曲线，线条简洁而明快，不过于雕琢、堆砌。旗袍的造型变化十分微妙，衣身的长短、开衩的高低及袖子的有无，这种简单的线条变化，或传达出玲珑的气质，或显示出丰腴柔美，在掩盖身材缺陷的同时，又恰到其分地凸显东方女性的曲线之美。

The beauty of the cheongsam comes to be unique and full of classical spirituality, which gets well integrated with the body and temperament of the wearer to show the gentleness, gracefulness, and sensuality of women perfectly, showing an oriental subtle style. Cheongsam can naturally outline the curves of women's bodies with simple lines and proper design. Changes in this costume's styles vary subtly in the length, openings, and sleeves. While, such simple changes in lines contribute a lot to hiding the defects of the figure and highlighting the beauty of oriental women's curves properly, thereby conveying their diverse temperaments.

2. 婉约之美/The Beauty of Grace

旗袍又是含蓄的，她宛如女性的第二层皮肤，无须过多装饰，简洁的线条便能将女性曲美的身形显露出来。旗袍也是张扬的，一条从脚踝开到腿部的高衩，恰如其分地显现了女性胴体的曲线美，动起来是一道令人陶醉的风景，静下来则如一幅耐人寻味的画卷。一动一静间增添了无限奇妙的想象空间。旗袍的曲线是对人体的一种温和释放，显现出健康而不媚俗的气质。旗袍看似包裹密实，但不经意间展现了所有能展示的，把东方女性的含蓄、内敛表达得淋漓尽致，处处显得精致、飘逸、含蓄，让人越发觉得高贵与典雅。

The cheongsam is implicit, like the second skin of a woman, able to reveal her beautiful figure with proper decoration and simple lines. The cheongsam is also flamboyant. Its high slits from the ankle to the leg appropriately demonstrate the curvaceous beauty of the female carcass. It is a picturesque scene when it moves, and an intriguing picture when it stays still, leaving infinite space for wonderful imagination. The cheongsam's curve is a relatively gentle release to the human body, showing a healthy temperament. Despite its seemly tight image, cheongsam gives a full expression of the implicit and restraint spirit of oriental women inadvertently, making them more elegant and noble.

3. 点睛之美/Beauty as Highlight

旗袍挺拔的外形、柔和的气质、文雅的品格，显现出简约的典雅与高贵的单纯，给人留下深刻的印象。除此之外，那画龙点睛之笔必定是盘扣。盘扣作为一种传统的服饰手段，极具中国传统民族风情，是中华民族文化的美好象征。在旗袍的衣身上精工再造，令人叹为观止。盘扣是中国人的精致手艺，盘扣造型背后蕴含的中华民族的文化内涵，是中华民族的魅力的重要体现。盘扣以画龙点睛的方式出现在当代旗袍的设计上，既传达了现代人的设计理念，又与现代服饰完美结合，走上国际舞台，使旗袍更具民族风情。

The cheongsam's tall and straight appearance, soft temperament, and elegant character combine into simple grace and noble simplicity, leaving a very deep impression on people. And the Chinese knot button must be another highlight. As a traditional means of clothing, the Chinese knot button is a decoration with typical traditional Chinese national customs and a beautiful symbol of the Chinese national culture. Its recreation on the cheongsam is amazing. Standing for the exquisite craftsmanship of the Chinese, the knot button contains the cultural connotation of the Chinese nation, and it plays an important role in the design of contemporary cheongsam to convey the design concept of modern people. Its perfect combination with modern clothing on the international stage allows the cheongsam to be more ethnic.

第二节　　旗袍结构设计/Cheongsam Structure Design

一、传统旗袍结构/Structure of Traditional Cheongsam

传统旗袍在结构设计上，多采用直线条和宽松的造型，除了领口部位略微呈现出立体状态外，其肩部、胸部、腰部和臀部等均为平面构成形式，几乎不存在立体结构感。因此，很难从衣身和袖子的造型中看出人体曲面各部位的变化、人体前后的体面差异。这种只重视细节刻画而相对忽视整体塑造的特点，在传统旗袍的造型中显得尤为突出。传统旗袍基本上是一个平面结构、上下连成一体的服装款式。

In terms of structural design, traditional cheongsam mostly adopts straight lines and loose shapes. Except for the collar, which is slightly three–dimensional, its shoulders, chest, waist and buttocks are all flat, with little sense of three–dimensional structure. It is difficult to see changes in various parts of the human body's curved surface and the decent gap between the front and rear of the human body from the costume shape and sleeves. Such characteristic that only pays attention to details and ignores the overall shape features prominently in the style of traditional cheongsam. Therefore, the traditional cheongsam is a kind of one–piece clothing with a flat structure.

（一）传统旗袍衣身结构/Traditional Cheongsam Body Structure

传统旗袍的衣身宽松，线条平直，腋下略有收束，大襟右衽，胸围与下摆的宽度相近，呈平面直筒式造型。其造型上多以门襟的式样变化进行区分，有琵琶襟、如意襟、斜襟等几种款式。相较于先前的袍服，传统旗袍的衣身与人体间的孔隙略小，呈严密、挺拔的封闭状态。

The traditional cheongsam is loose in overall body, straight in lines, and slightly tight in armpits, with a big front and lapel on the right side and the same width in bust and hem, showing a flat straight shape. The shapes are mostly distinguished by the style changes of the front of the costume（called Lapel）, such as the Pipa Lapel, the Ruyi Lapel, and the Slanting Lapel. Compared with the previous gowns, the gap between the traditional cheongsam body and the human body has been narrowed down to a tight, straight and closed shape.

民国初期，女子的服饰主要以袄、裙、裤装为主，到了20世纪20年代以后，由长马甲和短袄合并衍生而出的服装，才是史学界所公认的"旗袍"，但其在结构上并未完全脱离传统的古典特质。这一时期的旗袍，其结构以布幅为准绳，恪守着中华传统服饰"十字形平面直线结构"的框架，即手臂平伸后与身体的直线成垂直交叉，整体设计宽松流畅，不显锋芒。

In the early years of the Republic of China, women's clothing was mainly jackets（Ao）, skirts（Qun）, and trousers（Ku）. After the 1920s, the cheongsam derived from the combination of long vests and short jackets was the "cheongsam" recognized by historians. However, its structure remained the traditional classical character. The structure of the cheongsam in this period was based on the fabric width, and adhered to the framework of the "cross-shaped plane and straight line structure" of traditional Chinese clothing, that is, the arms crossed the straight line of the body vertically after they were stretched flat to be loose and smooth without showing their sharp edges.

平面结构裁制，是以前后中线为基准，衣长方向为直丝，衣片中缝、前门襟中缝均为无缝对折，有时因布幅宽度的限制，不得不使用布边进行拼缝。侧缝保持直线形向外扩张，无腰部曲线结构。尺寸通常通过直接测量，采用定寸法或凭借经验来绘制结构图。由于宽松的胸腰围尺寸穿在身上，腋下部位会堆满衣褶，但这一结构所具备的却是一种若隐若现、摇曳生姿的美感。

The plane structure cutting, that is, the clothing is cut with the front and rear center lines as the benchmark, the length of the garment is straight, and the center seam of the garment piece and the center seam of the front placket are all folded in half seamlessly. The side seams flare out in a straight line without a waist curve. Dimensions are drawn by direct measurement, by sizing or by experience. The loose bust and waist size are worn on the body, and the underarms will be full of pleats, but this structure has a looming, swaying beauty.

（二）传统旗袍领子结构/Traditional Cheongsam Collar Structure

倒大袖旗袍时期的领型为立领。虽然传统旗袍领型中也存在圆领口的款式，也就是在衣身上不装领子，但是清末流行的低立领，以及后来逐渐提高的元宝领，几乎可以遮住人的半个脸颊，展现出女子纤细安静的气质，这些都是传统旗袍领型中的经典款式。

The collar style of the big sleeve cheongsam period was a stand-up collar. There was also the round collar style（referring to the clothing with no color on it）in the traditional cheongsam collar, in addition to other classical styles including the low stand–up collar popular in the late Qing Dynasty, the improved one on this basis in the late Qing Dynasty that can almost cover half of the wearer's cheek, and the Yuanbao–shape collar（looking like the golden ingot in ancient China）.

（三）传统旗袍袖子结构/Traditional Cheongsam Sleeve Structure

在传统旗袍制作中，左边袖子与左前或前大襟、左后片连裁成一幅，右边袖子与里襟连裁成一幅，门襟独立裁剪，袖子没有肩缝。后身左右两侧及里襟采用三角形插片进行拼接，这主要是为了节省面料。袖长依据布幅而定，袖口的拼接是为了满足袖长尺寸，接袖部分有刺绣装饰，更能实现耐穿性，可谓"敬物尚俭"的智慧体现，如图2–21所示。袖口与袖根尺寸相等，或设计成倒大形喇叭状，呈直线形由衣身展出（原身出袖）。

In the traditional cheongsam, the left sleeve was cut into one piece with the left front or the front lapel and the left back piece, the right sleeve and the inner lapel were cut into one piece, with lapel cut independently, and the sleeves have no shoulder seams. The left and right sides of the back body and the inner lapel were spliced with triangular inserts to effectively save fabrics. The sleeve length was determined by the fabric width. The cuffs were stitched to meet the sleeve length. The sleeves were decorated with embroidery, making the clothing more durable, as shown in Figure 2–21. The cuffs were the same size as the cuffs, or in the shape of an inverted large trumpet, displayed in a straight line from the body（Grown–On Sleeve）.

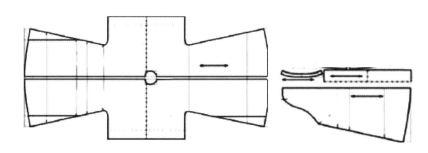

图2–21 受布幅所限传统旗袍的拼接
The Stitching of Traditional Cheongsam Limited by the Fabric Width

（四）传统旗袍整体结构设计/The Overall Structural Design of Traditional Cheongsam

在进行传统旗袍整体结构设计时，首先确定前、后衣片中心线和袖子中心线，两线在中点处十字交叉。前后衣片中心线的长度为衣长的两倍，一般从领口到离脚面2～3cm处；袖子中心线的长度为人站立时，两手臂伸直成一线，从左手腕到右手腕的距离；袖口线宽度为袖肥的两倍，袖口线可根据款式变化画成其他相应线形；前胸和后背的宽度相同，都是胸围的一半；前后下摆宽度一致，下摆宽度比胸围宽度略大，下摆线略微画出弧度；然后画出摆缝线和袖缝线，这两线用弧线连接；在前、后衣片中心线和袖子中心线交叉处画出领口弧线，注意前领深大于后领深；然后在前衣片右边画出开襟弧线，并画出里襟，至此传统旗袍的结构设计基本完成，如图2–22所示。整个衣片的设计以前、后衣片中心线为左右对称（开襟除外），以袖子中心线为前后对称（领口深浅除外）。前胸与后背在结构上没有差异，既不强调后背的挺拔也不突出前胸的曲线，前后衣片结构除开襟设计和领口深浅外没有区别。

When it comes to the overall structure of the traditional cheongsam, the center line of the front and rear parts and the center line of the sleeve should be first determined, and the two lines cross at the midpoint. The length of the center line of the front and rear parts is twice that of the clothing which is generally measured from the neckline to 2–3cm away from the feet; the length of the center line of the sleeves is the distance from the left wrist to the right wrist by standing and stretching out the two arms in a straight line; the width of the cuff line is twice the cuff fat, and the cuff line can be drawn into other corresponding lines according to the changed styles; the width of the front chest and the back are the same, being 1/2 of the bust; the width of the front and rear hem is the same, and the width of the hem is slightly larger than the bust width, and the hem line is slightly curved; then draw the hem line and the sleeve stitch line which are connected by an arc, draw at the intersection of the center line of the front and rear parts and the center line of the sleeve the arc of the neckline, noting that the depth of the front collar is greater than the that of the back one; draw the arc of the open Lapel on the right side of the front piece, and draw the inner Lapel. Then, the structural design of the traditional cheongsam is basically completed, as shown in Figure 2–22. The entire garment piece is symmetrical with the center line of the front and back parts (except for the open Lapel), and with the center line of the sleeves (except for the depth of the neckline). The curve of the front chest and the front and back parts are the same except for the design of lapel and the depth of the neckline.

传统旗袍在整体的结构设计上并不涉及腰部和臀部的尺寸，也不存在腰围线、臀围线，以及由腰臀差量形成的侧缝弧线变化。由此可以看出，传统旗袍在结构设计上并不关注、或不表现人体的曲线与曲面，而是一种直线与平面的宽松设计。

The overall structural design of the traditional cheongsam didn't consider the size of the waist and buttocks, with no waistline, hipline, and side seam arc changes formed by the difference

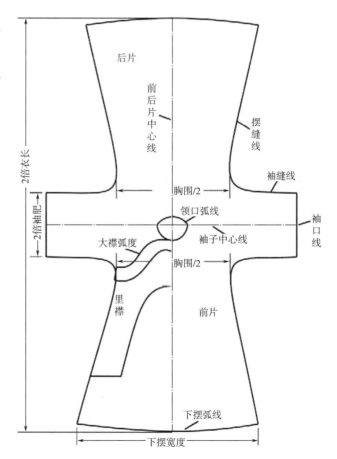

图2-22　传统旗袍结构图
Structure of Traditional Cheongsam

between the waist and hips. In this sense, the traditional cheongsam structure design paid little attention to curves and curved surfaces of the human body, being a straight, plane and loose design.

二、改良旗袍结构/Structure of Reformed Cheongsam

如上所述，传统旗袍造型平稳而单纯，主要采用直线条设计。除了领口部位略微表现出立体形态外，肩、胸、腰及臀部皆呈现平面化的态势，因此很难从衣身和袖子的造型中看出人体曲面的变化，如人体前后的体面差及手臂立体状态的平面体现等。但由于旗袍在演变过程中不断吸收西式立体裁剪的方法理念，现代旗袍的造型已能够较好地体现人体的体面结构，被誉为中华服饰文化的杰出代表，其造型线条简练、优美大方，穿着时与人体体态贴合度较高，能够展现出人体的曲线美。旗袍的结构设计主要经历了以下几个阶段。

As mentioned above, the traditional cheongsam was one-piece and simple in shape, mainly designed with straight lines. The three-dimensional shape can only slightly be seen in the collar, excluding shoulders, chest, waist and buttocks which were all flat. Hence, it is difficult to see the changes in the human body surface from the shape of the garment and sleeves, such as the front and rear of the human body, as well as the plane view of the arms' three-dimensional state in standing condition, etc. However, the shape of modern cheongsam, due to the continuous absorption of Western-style three-dimensional cutting methods and concepts throughout the evolution process, can do a good job of presenting the human body in a stereoscopic way, thereby being praised as an outstanding representative of Chinese clothing culture. It has a high degree of fit with the human body, showing curvaceous beauty with simple lines and elegant design. The structural design of cheongsam went through the following stages.

（一）20世纪30～40年代旗袍过渡时期的结构设计/The Structural Design of the Cheongsam during the Transitional Period from the 1930s to the 1940s

19世纪20年代末30年代初，新式旗袍应运而生，它吸收了西式裁剪方法，特点是腰身较宽松，但已有明显的廓型，侧缝采用弧线裁剪；衣长至脚面或小腿，袖口宽大多变，并做绲边镶边，如图2-23所示。30年代中后期，旗袍袖口逐渐缩小，衣长至膝盖下，腰身开始合体，从此，直观暴露的审美观逐渐取代了传统含蓄隐晦的审美意识。衣片前后中心、肩缝仍然是连裁，如果是长袖旗袍，则下半部分是断裁的。胸围、腰围、臀围放松量相对较大，腋下取一胸省，腰部相对自然宽松，两侧开衩及领高适中。结构设计的主要难点是袖笼深，采用定数或通过胸围的比例计算，如按胸围的2/10计算挂肩的长度，如图2-24所示。

In the late 1820s and early 1930s, the new cheongsam that absorbed the Western cutting method came into being, characterized by a loose waist, blurred silhouette, and the side seam arc cutting; the length of the garment reached the top of the feet or the calf, the width of the cuffs varies, and the hem is trimmed, as shown in Figure 2-23. In the mid-to-late 1930s, the cuffs of the cheongsam got narrowed, the length of the dress reached below the knees, and the waist began to fit. This marked the replacement of the traditional implicit and reserved aesthetics by the intuitive and exposed aesthetics awareness gradually. The front and rear center and shoulder seams of the garment were still cut in one piece, and separated cutting was used for long cheongsam. The bust, waist, and hip relaxed width were relatively loose, with one chest dart under the armpit, loose waist width, openings on both sides, and moderate collar height. The primary difficulty of such a structure lies in the depth of the arm size, which was calculated by a fixed number or by the ratio of the bust, for example, calculating the length of the armhole based on 2/10 of the bust, as shown in Figure 2–24.

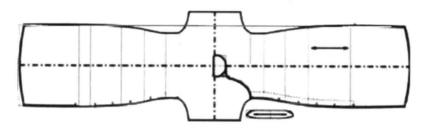

图2-23　过渡时期的旗袍廓型
Cheongsam Silhouette during the Transitional Period

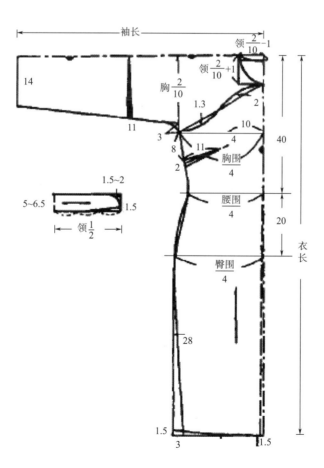

图2-24　过渡时期的旗袍结构设计
The Structure Design of the Cheongsam during the Transitional Period

（二）20世纪50~70年代旗袍的立体结构设计/The Three-dimensional Structure Design of the Cheongsam from the 1950s to the 1970s

20 世纪 50 年代中叶至 70 年代是旗袍走向立体结构的时期。1955 年 *McCALL'S PATTERN BOOK & McCALL'S NEEDLEWORK* 中记录的具有中式特征（立领、衽势、绲边）的西式连衣裙，采用了与《民国旗袍与海派文化》一文所定义"改良旗袍"相一致的"分身、分袖、施省"立体结构。这个时期的旗袍破开了肩缝，使腋下垂褶明显减少；增加了胸省和腰省，使胸型突显；装袖结构的出现，使肩部合体度增加。改良旗袍的完整形态符合立体造型需求的分身、分袖、施省结构，标志着它对传统古典旗袍和过渡时期旗袍"十字形平面结构"的彻底颠覆。无袖改良旗袍的主体结构由前衣片、后衣片、领片、小襟四部分构成，前片有胸省、腋下省及腰省，后片有腰省，肩缝破开，无袖，腰线有明显收腰设计，下摆内敛，如图2-25所示。

The mid–1950s to the 1970s was the period when the cheongsam moved towards a three–dimensional structure. As recorded in *McCALL'S PATTERN BOOK & McCALL'S NEEDLEWORK* in 1955, the "Western–style dress" with Chinese characteristics（stand–up collar, lapel, and piping）adopted the same three–dimensional structure of "separated clothing, separated sleeves, and darts" with the "reformed cheongsam", as defined in the *Cheongsam of the Republic of China and Shanghai-style Culture.* The cheongsam of this period broke the shoulder seam, which significantly reduced the drape under the armpit; increased the chest and waist darts, making the chest prominent; the set–in sleeve structure made the shoulder part more fit with body. The complete shape of the improved cheongsam conformed to the three–dimensional modeling needs of the clothing body, the separated sleeves, and the dart, which marked its complete subversion of the traditional classical cheongsam and its "cross–shaped plane structure" in the transition period. The main structure of the sleeveless improved cheongsam consisted of four parts: front piece, back piece, collar piece and small Lapel, with chest darts, underarm darts and waist darts in the front piece, and waist darts in the back piece; the shoulder seam got cut with no sleeves; there was an obvious design to narrow down the waistline, with restrained hem, as shown in Figure 2–25.

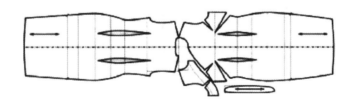

图2-25　无袖改良旗袍的廓型
The Outline of the Improved Sleeveless Cheongsam

符合立体造型需求的施省结构在这个时期完全成熟，整身共计4对8个省，以前后中线对称分布于衣身左右两侧，省量分配根据胸腰差和臀腰差的实际情况，形成省量多少的具体数值。加上大幅收摆设计，形成侧缝曲线最盛的造型，旗袍至此完全颠覆了古典和过渡旗袍的"十字形平面结构"传统，改良旗袍的立体结构与"布幅决定形态"的传统观念结合，展示了一个全新的华服时代的风貌，如图2-26所示。

The dart structure that meets the needs of three-dimensional modeling got fully mature in this period. There were a total of 4 pairs of 8 darts of the clothing which were symmetrically distributed on the left and right sides of the clothing based on the front and rear midlines, and the specific value of the darts was calculated according to the actual chest-waist and chest-hip difference. Coupled with the design of the large slung hemline, the most prominent side seam curve was formed, which marked the complete subversion of the "cross-shaped plane structure" tradition of the classical and transitional cheongsam. Therefore, it should be said that the integration of the improved cheongsam's three-dimensional structure with the traditional concept of "the cloth size determining the shape" unlocked a brand new era of Chinese clothing, as shown in Figure 2-26.

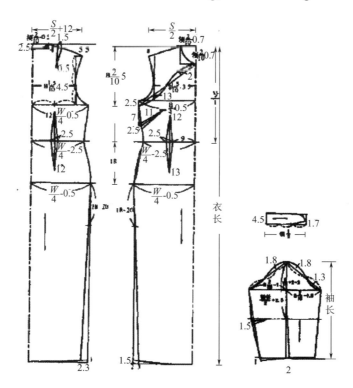

图2-26　改良旗袍的结构设计
The Structure Design of the Improved Cheongsam

这一时期旗袍造型已开始逐渐展示人体的柔美曲线，同时隐含了某种欲现还掩的小家碧玉的韵味。在造型结构变化方面，已由平面结构逐渐向立体结构过渡（即省道的形成与变化）。但是20世纪40年代的旗袍在结构设计上仍明显地存在一些不完全适体的部位，如肩斜、袖窿、袖片等部位还没有和具体的立体人形紧密地结合起来。其构成方法主要是通过精确的测量和通过比例构成法进行各部位的计算，后片肩省的运用符合人体肩胛骨立体造型。通过对该时期多款旗袍的复制研究可以发现以下特点。

The shape of the cheongsam during this period began to show the soft curves of the human body, and implied a charm of half-exposure and half-covering. In terms of the change of the modeling and structure, it gradually transitioned from a plane structure to a three-dimensional structure (for example, the formation and change of dart). However, the cheongsam in the 1940s still had some parts that remained unfit in the structural design, such as the shoulders, armholes, sleeves and other parts that failed to be combined with the specific three-dimensional human figures. Its composition method was based on accurate measurement and calculation of each part through the proportional composition method. The use of the dart in the shoulders at the back conformed to the three-dimensional shape of the human shoulder bones. From the copy and research of various cheongsams of this period, the following characteristics can be found.

（1）衣片、袖片和领片是分别裁制的，体现了立体的造型观念。

The garment piece, sleeve piece and collar piece were cut separately, reflecting its three-dimensional modeling concept.

（2）通过胸腰省、腋下省、后腰省和肩省的处理，来塑造女性胸、腰、臀三围曲线的性别特征和三位一体的曲线美。

The gender characteristics of women and the beauty of their curves in the three parts of chest, waist and hip were highlighted by adding darts in the chest and waist, armpit, back, and shoulders.

（3）旗袍的肩缝与前后中心线之间的角度小于90°（不像清代旗袍，肩线与前后中心线之间呈垂直状态），减少了手臂下垂时的腋下多余量。

The angle between the shoulder seam of the cheongsam and the front and rear center lines was less than 90° (unlike the cheongsam of the Qing Dynasty, the shoulder lines were vertical with the front and rear center lines), which made the underarms tighter when the arms are sagging.

（4）袖片的结构采用了一片合体式制图，前中线偏前2cm和袖肘处收省，更符合手臂运动和造型的美观。但是在袖宽是以胸围的2/10为计算基础或仅凭经验，缺乏科学性。

The structure of the sleeve piece adopted a one-piece integrated drawing, the front midline was 2cm forward and the sleeve elbow was narrowed, which conformed more to the beauty of arm movement and shape. However, the sleeve width was calculated on the 2/10 formula of the bust or based on experience, which was not scientific.

三、现代旗袍结构/Modern Cheongsam Structure

（一）20世纪80年代以后现代旗袍的原型法结构设计/Prototype Structure Design of Modern Cheongsam after the 1980s

现代旗袍无论从其外形塑造还是内部结构来看，都蕴含着东西方服饰的双重特点：袍服的基础形制源自中国传统服饰，而紧身贴体的外形强调形体的存在，表现了现代女性体态的曲线与柔美，再一次展示了中西合璧的现代旗袍无限的风采和时尚感。

The modern cheongsam, in terms of its appearance and internal structure, contains the dual characteristics of Eastern and Western clothing: the basic shape of the clothing is derived from traditional Chinese clothing, while the close-fitting shape emphasizes the beauty of the human body, showing the curves and softness of the modern female body as well as the infinite elegance and fashion sense of the costume that integrates the east and the west.

中西合璧的现代旗袍的造型特色是立体的、具象的、写实的，注重三维空间中的形体塑造，注重用显露甚至夸张的手法来突出表现人体美。从构成上看是曲线裁剪，衣、袖片分离，局部夸张，即裁片按照人体结构分别裁剪，大多有弧线；领型变化丰富，直观暴露的审美观逐渐取代了传统含蓄、隐晦的审美意识。20世纪80年代以后的旗袍是一个完全体现立体面的结构造型，原型法结构设计的出现，使旗袍穿着时更加贴体、合理、方便，从而能更自如地表现现代女性人体曲线与旗袍结构的吻合性，以突出纤腰、美臀、丰胸的成熟美，真正使旗袍款式逐步走向完善与成熟，如图2-27所示。

The modern cheongsam that integrates the east and west clothing features is three-dimensional, figurative and realistic, focusing on the shaping of the body in three-dimensional space to highlight the beauty of human body with revealing and even exaggerating techniques. In terms of composition, it is a curve cut, that is, the clothes and sleeves are separated and partially exaggerated. In other words, the pieces are cut according to the human body structure, and most of them have arcs with changeable collar shapes. The traditional implicit and obscure aesthetics concept has been gradually replaced by the intuitive and exposed aesthetics preference. The cheongsam after the 1980s was a completely three-dimensional structure model, and the birth of the prototype structure design made the cheongsam more suitable, reasonable and convenient to wear, so that it can express the conformity between the curve of the modern woman's body and the structure of the cheongsam much better. To highlight the mature beauty of a slim waist, beautiful buttocks, and plump breasts, the cheongsam style has gradually become perfect and mature, as shown in Figure 2-27.

通过对旗袍的复制研究可以发现原型法结构设计的以下特点。

By duplicating and conducting research of cheongsam, the following characteristics of the structural design of the prototype method can be found.

（1）从胸围的放松量看更加合体，通过对原型胸围的前片减小1cm，后片减小2cm，共减小6cm，使胸围的放松量为4cm，同时腰围、臀围放松量也保持在4cm。

From the perspective of the relaxation of the bust, it is more fit. By reducing the front piece of the prototype bust by 1 cm and the rear piece by 2 cm, the total reduction reaches 6 cm, so that the relaxation of the bust is 4 cm, while that of the waist and hip are kept at 4 cm.

（2）通过胸省、腋下省和腰省的微妙处理，绘制成枣核形，更加衬托出胸部丰满与性感，塑造女性胸、腰、臀三位一体的曲线美。

The delicate treatment of darts in the chest, underarm and waist leads to a jujube shape that can highlight the plumpness and sexiness of the chest to outline the curvaceous beauty of women's breast, waist and buttocks.

（3）旗袍三围尺寸的前、后片调节。为使前片更好地衬托胸部造型，侧缝线位于人体中

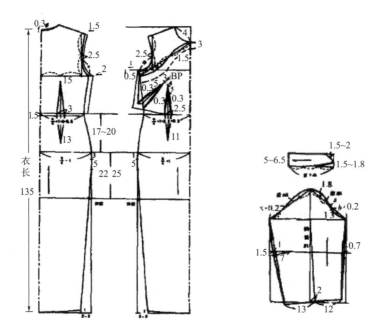

图2-27　现代旗袍的结构设计示意
The Structure Design of Modern Cheongsam

间，使前片胸围、腰围、臀围大于后片，前片在原型上减1cm，后片在原型上减2cm，即前片比后片大2cm（腰围、臀围相同）。

Adjustment of the front and back pieces of the cheongsam's measurements. In order to make the front piece better set out the shape of the chest, the side stitches are located in the middle of the human body, allowing the bust, waist, and hip circumferences of the front piece to be larger than those of the back piece, thereby reducing the front piece on the prototype by 1 cm, and the back piece by 2 cm, that is, the front piece is 2 cm larger than the back piece（the waist and hip circumferences remain the same）.

（4）袖片构成采用袖笼弧长的计算方法，首先分别测量出前、后袖笼弧长AH，选择适合的袖山高形式，AH/4+2.5cm（原型袖山高），然后分别通过前、后袖笼弧长AH画出前后袖宽，保证袖笼与袖山弧线的良好配合。

The sleeve piece is calculated based on the sleeve arc length, in which, the front and rear sleeve arc lengths AH are firstly measured respectively, and then the appropriate sleeve crown height is selected to get the AH/4+ 2.5 cm（prototype sleeve crown height）, and then the front and rear sleeves width can be drawn according to the arc length AH of the front and rear sleeves, ensuring good match between the sleeves and sleeve cap curve.

纵观旗袍造型的演变、发展可以看到，随着旗袍造型的每一次演变，作为服装造型设计与工艺之间的桥梁和手段，结构设计也在不断地变换。从平面到立体、从直线到曲线、从以衣为本到以人为本、从宽松到紧身的形制不停地变换，每一次变化都代表了一个特定历史背景下的文化审美倾向、时代精神及社会生产力与技术水平。

Throughout the evolution and development of the cheongsam shape, it can be seen that, as a bridge and means between clothing design and craftsmanship, the structural design changes with the clothing's shape. And each change in the structural design of modern cheongsam, which is from plane to three-dimensional, or from straight line to curve, or from clothing-oriented to people-oriented, or from loose to tight-fitting, represents the cultural aesthetic tendency, the spirit of the times and the level of social productivity and technology of a specific historical period.

旗袍的造型从平面造型、直线裁剪、强调工艺修饰为特色向以立体造型、曲线裁剪、表现形体美为主流的方面发展。经过不同年代的一次又一次变革，旗袍造型与人体更加紧密结合，自然流畅、简洁充分地展现女性的曲线美和多元化的美。旗袍的外形构成符合中国人的衣着要求和审美情趣，既体现中国传统文化，又具有现代的形式美感。可以说，旗袍是对中国传统服饰艺术的概括，它也为众多服装设计师提供了取之不尽的灵感源泉。

The shape of the cheongsam has undergone a process from a flat shape with straight line cutting and a focus on craftsmanship to a three-dimensional shape with curve cutting with a focus on body beauty. After reforms in different ages, it got closely integrated with the human body, naturally, fluently, concisely and fully showing the curvaceous and diversified beauty of the female. The shape and composition of the cheongsam not only conform to the clothing requirements and aesthetic tastes of the Chinese people, which not only reflect the traditional Chinese cultural history but also have a modern aesthetic sense. In this sense, cheongsam is a summary of traditional Chinese clothing art, provides exhausting inspiration for a large number of designers.

（二）旗袍省的结构设计/Structural Design of Cheongsam Darts

在服装的结构设计中，省的设置是决定服装造型是否美观、结构设计是否合理的关键因素之一。旗袍作为丝绸服装的典型代表，在追求线条简洁流畅的同时，也要求服装本身与人体具有较高的贴合度，力求体现女性的曲线美。因此，旗袍省的设置就变得格外重要，直接决定了旗袍与人体的贴合度及其造型的美观度。当然，作为紧身型服装，旗袍在合乎人体曲线及追求美观造型方面，还应考虑诸如领部起翘高度与领外缘线的造型关系、袖型结构与其外观造型关系、下摆与开衩的造型关系等方面的问题。作为贴身的上下一体的旗袍来说，影响整个造型效果的腰身处的省量分配与侧缝线的形态关系，对其造型的美观性有着重要的影响。以一款无袖经典款式的旗袍作为范例，在其纸样的形成过程中，对其胸、腰省处空间省量的分配与侧缝线的形态关系进行研究，以寻求一种比较好的省量分配关系，使侧缝线的形态更美观。

In the structural design of clothing, the design of a dart is one of the key factors that determine the clothing's shape and structure. As a typical representative of silk clothing, cheongsam, while pursuing simple and smooth lines, also requires the clothing to perfectly fit with the human body, so as to reflect the beauty of women's curves. Therefore, the design of dart in cheongsam, which plays a decisive role in achieving a perfect fit with the human body and the beauty of its shape, becomes particularly important. Of course, in addition to these, the cheongsam, as a tight-fitting garment, should also consider other issues, for example, the relationship between the height of the collar and the outer edge of the collar, the relationship between the sleeve structure and its appearance, and the shape of the hem and openings, etc. For cheongsam, a close–fitting, one–piece dress, the relationship between the dart distribution at the waist and the shape of the side stitches that directly affects the entire modeling effect, has a great implication for the aesthetics of its modeling. Taking a sleeveless classic cheongsam as an example, in–depth research on the relationship between the distribution of darts in the chest and waist parts and the shape of the side seam helps a lot to achieve a beautiful side seam in the process of making.

范例研究时选择的三围分别为胸围84cm、腰围64cm、臀围91cm。纸样选择以旧文化式原型为基础，用平面裁剪方法绘制旗袍的基本纸样并修正，如图2-28所示；以修正好的基本纸样为基础，仅改变纸样的前、后片省量分配比例关系，绘制旗袍纸样并补正，以确保其他部分的一致性，如图2-29所示。在补正的过程中，要特别注意对每一个省道甄别确定，同时注意省道之间的相互协调关系。

The measurements chosen in the case study were 84cm for the bust, 64cm for the waist, and 91cm for the hip. The pattern selection was based on the old cultural prototype, and the basic pattern of the cheongsam was drawn and corrected by the flat cutting method, as shown in Figure 2–28; based on the corrected basic pattern, Figure 2–29 was completed and corrected by only changing the dart proportion and distribution of the front and rear pieces to ensure the consistency of other parts. In the process of correction, special attention should be paid to the identification and determination of each dart, and at the same time, as well as the mutual coordination relationship in between.

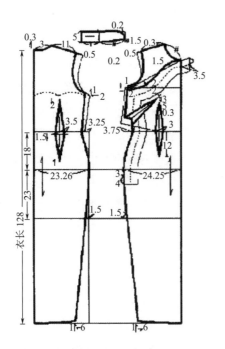

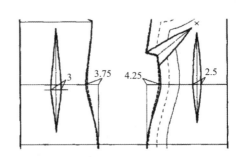

图2-28　补正前基本1#纸样
Basic 1# Pattern before Correction

图2-29　补正后腰部放大的2#纸样
The Enlarged 2# Pattern of the Waist after Correction

有5种不同的前、后衣片省量分配比的纸样变化及补正，以补正后的旗袍基础纸样为参考，做前、后片省量分配比的变化处理。在衣身前、后片省量分配处于不同的比例关系时，分别有后腰省（记作省 1）、后身片侧缝线处省（记作省 2）、前身片侧缝线处省（记作省 3）、前腰省（记作省 4）间搭配不同的几种情况。此外，考虑到更具体的制板规则，去掉一些在结构与造型方面不太合理的情况，最终确定以下 5 种情况，即 $W/4+ 1+0.25$（− 0.25），$W/4+ 1+ 0.5$（− 0.5），$W/4+ 1+ 0.75$（− 0.75），$W/4+ 1+ 1$（− 1），$W/4+ 1+1.5$（− 1.5）为研究对象。在基础纸样的基础上进行制图，并将它们分别记作 1#（基础纸样）、2#、3#、4#、5#。2# ~ 5# 纸样用丝绸面料缝制成样衣，请模特试穿、照相并补正、再照相，以获得满意的效果。由于丝绸面料具有较好的悬垂性能，对旗袍外观造型会产生一定影响。为了减少面料因素对服装造型产生的影响，在旗袍衣长上采用短款数据，从而提高实验的准确度。将补正后的样衣拆片、拓板，得到修正后 2# ~ 5# 的旗袍纸样，如图 2-30 所示。其中，实线为补正前的轮廓线，虚线为补正后的轮廓线。

There are 5 different pattern changes and corrections for the distribution ratio of darts in the front and rear parts. Take the corrected basic pattern of the cheongsam as a reference, and make changes to the distribution ratio of the dart in front and rear parts. When the dart distribution of the front and back parts of the body is in different proportions, it can be divided into the dart at the back waist (denoted as dart 1), the dart at the side seam of the back body (denoted as dart 2), the dart at the side seam of the front body (denoted as dart 3), and the dart at the back waist (denoted as dart 4). Given the more specific pattern making rules, some unreasonable situations in terms of structure and shape are removed, and then the following five situations are finally determined, namely taking $W/4+ 1+0.25$（− 0.25）, $W /4+ 1+ 0.5$（− 0.5）, $W /4+ 1+ 0.75$（− 0.75）, $W /4+ 1+ 1$（− 1）, $W /4+ 1+1.5$（− 1.5）as the studying objects. Draw on the basis of the basic pattern and record them as 1#（basic pattern）, 2#, 3#, 4#, and 5# respectively. Sew the 2# ~ 5# paper pattern with silk fabric to make the sample clothes, ask the model to try it on, take photos and make corrections, and then take photos to get satisfactory results. Because silk fabric has good drape performance, it will have a certain impact on the appearance of cheongsam. To reduce the influence of fabric factors on clothing shape, short data has been used on the length of cheongsam, thereby improving the accuracy of the experiment. Disassemble and rub the corrected sample garment to obtain the corrected 2# ~ 5# cheongsam pattern, as shown in Figure 2-30. The solid line is the contour line before correction, while the dotted line is the contour line after correction.

样衣 1# ~ 5# 主要是在旗袍空间省量的分配比例上进行变化，所以 5 件样衣所出现的不合体情况大致相同。总体来看，5 件样衣的补正主要集中在对侧缝线的处理，补正后的样衣皆在腰围处放出了 2cm 的量，而臀围值除了 2# 减少了 1.2cm 的量外，另外几件样衣皆减少了 0.6cm 的量。对补正后的样衣而言，胸腰差量减小，即它们各自前、后身衣片的总腰省量由原来的 10cm 变为了 8cm。在 5 件样衣各自的补正过程中，就空间省量的分配而言，只是改变了省 2 与省 3 的大小，但整个空间省量的分配比发生了变化。

Changes are only made for the samples 1# to 5# mainly in the distribution ratio of the

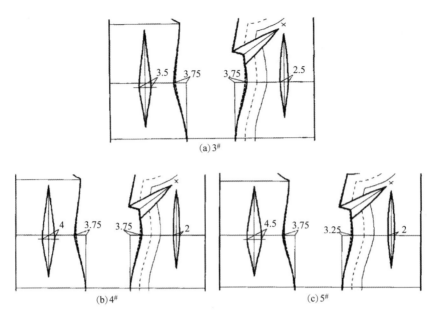

图2-30　补正后腰部放大的3#、4#、5#纸样
The Enlarged Samples 3#, 4#, 5# Pattern of the Waist after Correction

cheongsam's space darts, so the 5 samples do not fit in the same parts. The correction of the 5 samples mainly focuses on the treatment of the opposite side seamlines. The corrected cheongsam samples are all relaxed by 2 cm at their waists, and the hip circumference is reduced by 1.2cm except for 2#, while the remaining samples are reduced by 0.6 cm. For the corrected sample garments, it is equivalent to a reduction in the difference between the chest and waist, that is, the total waist darts of their respective front and rear body pieces have changed from the original 10cm to 8cm. During the correction process of each of the 5 samples, in terms of the distribution of space darts, only dart 2 and dart 3 have been changed, but the distribution ratio of the entire space darts remains unchanged.

　　从补正后样衣的外观效果来看，1#的空间塑型最好，侧缝线的形态也比较好，但在它稍微有些过于向前倾，整体效果属于一般偏好。2#的空间塑型较好，侧缝线的形态稍微有些曲度，在腰线部位有微小的前倾量，但其侧面状态产生了一些斜向的绺，影响了整件样衣的美观，总体而言，其适体度与美观性皆属于比较好。3#的情况与2#的情况相似，侧缝线的形态比较好，但美观性不够好。4#的空间塑型一般，侧缝线的形态接近直线，且其侧面产生的绺最多。另外，从侧面来看，胸下部位接近于平面，未能塑造出人体正面的曲面变化，在很大程度上影响美观，其适体度也不是很理想。5#的情况与4#相近，其侧缝线的形态已几乎是一条直线，且从侧面来看，该样衣不但未能塑造出人体的曲度，反而出现了前突的外观效果，所以其造型效果最差。此外，通过询问模特穿着过程中对各样衣的适体度的体验，得知4#的适体度最好，2#的适体度比较好，1#的适体度一般，3#与5#的适体度也皆属一般。综合外观效果和模特的适体度评价，可知2#的旗袍效果最好。

　　Judging from the appearance of the sample garment after the correction, Sample 1# has the best space shape, with a relatively beautiful shape of the side seam, but it feels a little too far forward, therefore its performance is regarded as above the average in terms of the overall effect. Sample 2# is good in space shape with a slightly curved shape of the side seam and a slight forward inclination at the waistline, but some slanting lines appear on the side which affects the aesthetics of the whole sample. In general, its fitness and aesthetics are relatively good. Sample 3# is similar to Sample 2#. The shape of the side seam is good, but it fails to show perfect aesthetic feeling in general. As for Sample 4#, it presents ordinary space shape, and its side seams are virtually straight with the most slanting lines on sides. Moreover, the lower part of the chest looks like a plane structure from the side. It fails to shape the surface changes of the front of the human body and to a large extent affects the appearance of the clothing, and its fitness is not ideal. Similar to Sample 4#, Sample 5# makes the shape of its side seam almost a straight line. From the side, the sample garment not only fails to shape the curvature of the human body, but even looks protruding in the front. Its shape effect is the worst. In addition, by asking the model how they feel about the fitness of various clothes, it is known that in her opinion that Sample 4# has the best fitness, followed by Samples 2# and 1#, and then Samples 3# and 5#. Based on the appearance effect and the fitness evaluation of the model, it is believed that the cheongsam of Sample 2# has the best effect.

由实验结果可知，当衣身前、后片省量的分配比为0.5∶0.5时，旗袍的空间省量分配与侧缝线形态间的搭配关系比较合理，此时坯布样衣的侧缝线形态较好；对于选定的模特来说，衣身后片省量、后侧缝线处省量、前侧缝线处省量及衣身前片所收取省量的比例关系分别为3/1/1.5/2.5。由于人体结构、面料性能等因素都会影响成衣的最终效果，在制图时，可以参考此结论并综合考虑影响旗袍穿着效果的诸多因素，最终确定前、后片省量的大小。

Through experiments, it is found that when the distribution ratio of the dart in front and rear pieces of the body is 0.5∶0.5, the relationship between the space dart distribution of the cheongsam and the shape of the side seam is more reasonable to achieve good side seam shape; for the selected model, the ratio of dart at the back of the body, the dart at the back side seams, the dart at the front side seams, and the dart at the front of the body is 3/1/1.5/2.5. Since the human body structure, fabric properties and other factors affect the final effect of a garment, this conclusion can be referred in combination with other factors that affect the clothing effect to determine the size of the dart in the front and rear pieces.

思考题

（1）现代服装应该如何借鉴旗袍元素进行设计？

（2）各个时期的旗袍衣身结构具体有哪些变化？

（3）旗袍哪些部位可以设置省道？省量如何变化？

（4）款式系列设计。

① 2022年春夏流行趋势研究（主题提案手册）。

要求：制作形式为PPT，手册所展示的各个部分应具有延续性，设计具有一定原创性和前瞻性，并成系列展开，效果图、款式系列渐变等绘制精细整洁，允许加入制作工整的实样或小样。文字部分必须电脑打印，字型样式变化根据手册和设计需要来选择。

② 主题理念展板。

要求：主题名称、主题理念文字、意境图，尺寸为A3。

③ 制服或生活类旗袍系列化设计。

要求：系列旗袍研究与设计、设计线稿不少于100款，彩色稿不少于30款，手绘。制作尺寸为A4，装订成册。

一个系列的制服或生活类旗袍设计。材料的选用、色彩的搭配、款式系列等设计效果与手册中的设计理念有很强的联系，充分体现明确的创意，并符合流行趋势。工艺的设计符合服装整体主题的风格。对于成衣，其结构、工艺的设计应符合企业生产要求和市场要求。

Questions

（1）How should modern clothing learn from the elements of cheongsam in design?

（2）What are the specific changes in the body structure of the cheongsam in various periods?

（3）Which parts of the cheongsam are appropriate to be designed with darts? How does the dart change?

（4）Style series design.

① 2022 spring/summer fashion trend research（theme proposal manual）.

Requirements: Make a portfolio in PPT to present original and forward-looking designs in series that can be further created in the future, with refined renderings, and paintings of the style evolution. Well-produced real samples and small samples are allowed to be added. The text part must be printed by computer, and the font style changes are subject to the manual and design needs.

② Themed idea board.

Requirements: theme name, idea text, mood map. The size is A3.

③ Serialized design of uniform or daily wear cheongsam.

Requirements: research and design of series of cheongsam, no less than 100 design line drafts, no less than 30 color drafts, hand-painted. The production size is A4, bound into a booklet.

A collection of uniform or daily wear cheongsam designs. There should be a strong relation between the selection of materials, color matching, style series and other design effects with the design concepts in the manual, which fully reflects the clear creativity and conforms to the fashion trend. The design of the craftsmanship shall be in line with the style of the overall theme of the clothing. For ready-made garments, the structure and process design should meet the production requirements and market requirements of the enterprise.

○ 第三章

旗袍匠艺之心
The Heart of Cheongsam Craftsmanship

第一节　旗袍排料裁剪/Cheongsam Pattern Cutting

一、旗袍排料裁剪的方法/The Method of Cheongsam Pattern Cutting

（一）传统旗袍/Traditional Cheongsam

设置传统旗袍是在面料幅宽最小状态下，即在仅能满足通袖长，不考虑衣长方向的用料。这种假想条件虽然可以解释前、后中拼缝的形成，但是无法解释插摆破缝的形成原因。当面料最小幅宽增加到77cm，虽未超过天然丝织物的最大可织造幅宽，但可解释插摆拼缝的原因，如图3-1所示。

The traditional cheongsam is based on the hypothesis that when it comes to the smallest fabric width that can only meet the sleeve length, the material used for the garment length shall not be considered. Such hypothetical conditions can explain the formation of the front and rear middle seams, but not the formation of the insertion and sewing seams. When the minimum width of the fabric is increased to 77cm, within the maximum weavable width of natural silk fabrics, the insertion and swing seams can be explained, as shown in Figure 3-1.

通过图示对比发现，其面料使用量从603cm缩减至466cm，节省了近23%，这个实验阐释了插脚的形成原因。相较于图3-1（a）的结果，图3-1（b）中面料幅宽≥77cm的实验环境更加接近真实面料的使用情况。通袖长尺寸为121cm的情况，在民国初期不发达的纺织工业生产背景下（多为手工织造），这个指标远超过一般天然丝织物的最大幅宽，因此，只有破开中缝，使用双幅拼接，才有可能完成。布幅作为主要限制条件，促使传统旗袍"中缝"结构的产生。

From the graphic comparison, it is found that the amount of fabric used is reduced from 603cm to 466cm, with nearly 23% of fabric saved. This experiment explains the reasons for the formation of pins. Compared with the results in Figure 3-1（a）, the experimental environment with fabric width ≥ 77cm in Figure 3-1（b）is closer to the actual use of fabrics in reality. In the case where the length of the sleeve from shoulder to wrist is 121cm, in the context of undeveloped textile industry production conditions in the early Republic of China (mostly hand-woven), this indicator far exceeded the maximum width of ordinary natural silk fabrics. Therefore, only when the midline is cut off can a cloth be made with double stitching method. As the main limitation, the fabric width has prompted the creation of the traditional cheongsam's "middle seam" structure.

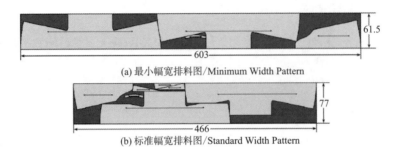

61.5

603

(a) 最小幅宽排料图/Minimum Width Pattern

77

466

(b) 标准幅宽排料图/Standard Width Pattern

图3-1　不同幅宽的旗袍排料分析
Pattern-cutting Analysis of Cheongsam with Different Widths

　　传统旗袍的典型特征——前后衣身的中缝，是区别第一和第二阶段的标志，但这一特征过去并未受到足够的关注。合格的裁缝师傅能够运用工艺手段将左、右两侧面料的花纹完美对接，使中缝不易觉察，所以在观察照片甚至是鉴赏实物时，中缝往往被忽略。1929年，民国政府正式颁布《制服条例》，该条例中女子礼服长衣的图版上清晰地画有中缝，明确标示出这一时期旗袍衣身中缝的正统性。

As the typical feature of traditional cheongsam, the center seam on the front and back of the clothing is a sign that distinguishes the first and second stages, but this feature has not received enough attention in the past. Qualified tailors could perfectly match the patterns on the left and right fabrics with craftsmanship to make the middle seam not easy to detect. That's why when observing photos or even appreciating real objects, the middle seam is often ignored. In 1929, the government of the Republic of China officially promulgated the *Regulations on Uniforms*, in which the middle seam was clearly drawn on the pattern of the women's dress gown, marking the orthodoxy of the middle seam of the cheongsam during this period.

（二）改良旗袍/Improved Cheongsam

　　改良旗袍侧缝收腰量不断增加，其整体廓型也由传统旗袍的宽松直线向相对合体的曲线转变，形成了"十字形平面曲线结构"。在结构上由于松量的减少，下摆收紧，使完整衣片可以容纳在1个布幅中，前、后片中间的破缝便消失。这个时期的旗袍为了使裁片能在一幅内裁剪，使用了"偷襟（也称拉襟或挖大襟）"裁法，这个方法极大限度地节省了材料，且延续了"十字形平面结构"的传统。其创新之处在于不破开肩缝、连身连袖的情况下，通过对面料的折叠和拔烫取得面襟与里襟的缝份及搭叠量，以完整的旗袍裁片（不需要破开前后中缝）坚守"十字形平面曲线结构"的中华传统。

The waistline of the improved cheongsam has been continuously narrowed, and its overall silhouette has also changed from the loose straight line of the traditional cheongsam to a relative shape that fits and highlights the curves of human body, forming a "cross-shaped plane curve structure". In terms of structure, due to the increasingly tight structure and hem, the complete garment piece can be accommodated in one cloth panel, and the middle seams in the front and back disappear. To make the pieces of cheongsam in this period cut in one piece, the cutting method of "Tou Lapel (also known as La Lapel or Wa Da Lapel, means)" was used, which greatly saved materials and passed on the tradition of "cross-shaped plane structure". Its innovation lies in that, given uncut shoulder seam and sleeves, the seam and overlap of the front and inner front can be obtained by folding and ironing the fabric, and the complete cheongsam piece (with no need to cut the middle seams in the front and back) has stayed true to the Chinese tradition of "cross-shaped plane curve structure".

步骤1：根据衣长确定所需面料的大小并画出横开领*AC*、直开领*BD*、前中心线*AE*和后中心线*BF*，如图3-2所示。

Step 1: Determine the size of the required fabric according to the length of the garment and draw the horizontal open collar *AC*, straight open collar *BD*, front centerline *AE* and rear centerline *BF*, as shown in Figure 3-2.

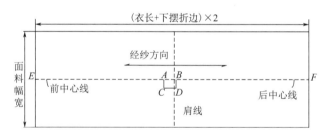

图3-2　偷襟步骤1
Step One for Tou Lapel

步骤2：如图3-3所示画出前后抬肩线、大襟线和荒裁线。

Step 2: As shown in Figure 3-3, draw the front and rear shoulder lines, the big Lapel line and the cutting line.

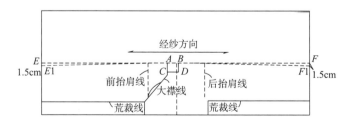

图3-3　偷襟步骤2
Step Two for Tou Lapel

步骤3：沿着前中心线*AE*和后中心线*BF*双折面料，并在图3-4中所示部位拔开，使*AE*1成为新的前中心线，*BF*1成为新的后中心。

Step 3：Double-fold the fabric along the front centerline *AE* and the back centerline *BF*, and pull it out at the position，as shown in Figure 3-4, making *AE*1 the new front centerline and *BF*1 the new back center.

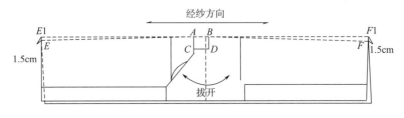

图 3-4　偷襟步骤 3
Step Three for Tou Lapel

步骤 4：沿着荒裁线将图 3-5 中灰色部分裁剪掉并剪开大襟。

Step 4：Cut out the gray part in Figure 3-5 along the cutting line and cut out the big lapel.

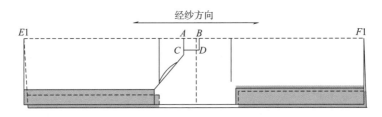

图 3-5　偷襟步骤 4
Step Four for Tou Lapel

步骤 5：如图 3-6 中所示方法使 *AE* 重新成为前中心线，*BF* 重新成为后中心线，最终得到了所需的缝份量，即阴影部分的搭叠量。

Step 5：As shown in Figure 3-6, make *AE* the front centerline again, and *BF* the rear centerline, and finally get the required seam weight, that is, the overlap amount of the shadow part.

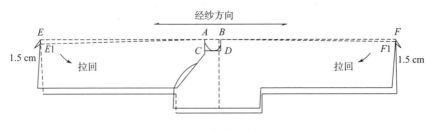

图 3-6　偷襟步骤 5
Step Five for Tou Lapel

改良旗袍通袖长 69.2cm ，是成衣尺寸的最宽处，也是布幅宽，如图 3-7 所示。通过这个时期的文献记录和裁剪图可知，此时的旗袍常依料而裁，通袖尺寸一般受布幅限制，通袖长一定小于或等于布幅宽度；而长袖旗袍则要求在布幅范围之外单独"拼袖"，达到"人以物为尺度"的境界。

The sleeve length of the improved cheongsam is 69.2cm, which is the widest part of the garment size, and it is also the width of the cloth, as shown in Figure 3–7. According to the documentary records and cutting drawings of this period, it is found that the cheongsam at this time was often cut according to the material, and the size of the sleeves is generally limited by the

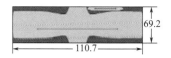

图3-7 改良旗袍排料分析
Pattern Analysis of Improved Cheongsam

width of the fabric and must be less than or equal to the width of the fabric width；while the long–sleeved cheongsam was required to stitch additionally beyond the fabric width, making it possible to take figure as the scale.

（三）立体旗袍/Three-dimensional Cheongsam

此时的旗袍，因为结构彻底立体化，裁片变得零散，且相应面料的种类更加丰富，幅宽已不构成限制因素。在这个前提的极限状态下，当面料幅宽≥臀围/2+2cm（缝份）≥50cm时，便不需要再增加前后中心线的剪开缝；当面料幅宽≥臀围+4cm（前、后衣片缝份）≥100cm时，仅需要1个衣长的面料便可完成裁剪，如图3-8所示。

Cheongsam at this time had already transformed into a three–dimensional structure, therefore, its pieces became scattered. And the fabrics for cheongsam have gotten increasingly abundant, making the width of fabric no longer a limitation. According to the limit of this premise, when the fabric width ≥ hip circumference/2 + 2cm（seam width）≥ 50cm, there was no need to increase the cutting and opening of the front and rear center lines; when the fabric width ≥ hip circumference + 4 cm（front and back seam allowances）≥ 100cm, only one garment length of fabric shall be needed to for cutting, as shown in Figure 3–8.

综上所述，在分身、分袖、施省的结构驱使下，臀围95cm为衣身围度尺寸最大处，根据实际裁剪的1/2计算，这个尺寸仅为47.5cm。对照这个时期的文献记录，臀围作为成衣围度的最大值，服装的结构完全摆脱了布幅的限制，无论是窄幅的锦缎还是宽幅的棉、麻、呢料，都可以顺利裁剪。

To sum up, driven by the structure of the separating cutting of clothing and sleeves, the hip circumference of 95cm became the largest size of the body circumference. According to the calculation of one–half of the actual cutting, this size shall only be 47.5cm. Based on the literature records of this period, the hip circumference was the maximum circumference of the garment, and the structure of the garment was completely free from the limitation of the cloth width. Materials ranging from the narrow brocade to the wide cotton, linen, and woolen can be smoothly cut.

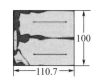

(a) 最小幅宽排料 / Fabric with Minimum Width

(b) 标准幅宽排料 / Fabric with Standard Width

图3-8 立体结构时期的旗袍排料分析
Pattern Analysis of Cheongsam during the Three–dimensional Structure Period

（四）旗袍各时期的裁剪技术总结 /Summary of Cheongsam's Cutting Techniques in Different Periods

1. 中式平面裁剪 /Chinese Flat Cutting

从裁剪角度看20世纪20年代的旗袍，都还保留着平面的裁剪结构。此阶段的旗袍没有肩缝，前、后衣片相连，衣袖和衣身相连。其基本裁剪步骤如下。

Viewing from the cutting, the cheongsam of the 1920s has still remained the characteristic of the flat cutting structure, with no shoulder seams, and the front and back parts were connected, so did sleeves and the clothing body. The basic cutting steps are as follows.

（1）先将面料沿 a 线水平对折，再沿 b 线垂直对折。

First, fold the fabric in half horizontally along line a, and then fold vertically in half along line b.

（2）在折好的面料上画出领口线、侧缝线、小腰、下摆线及袖口缝线，标出衣长、腰身、挂肩、下摆宽度、下摆起翘量，外缘线采用平直线条。

Draw the neckline, side seam, small waist, hem line and cuff line on the folded fabric, and mark the length, waist, armholes, hem width, and jut–out value of the hem. The boundary lines should be straight.

（3）在平面内空出1个领窝，开1条襟位线。

Leave a collar line in the plane and draw a line for Lapel.

（4）前、后衣片，左、右袖片连在一起裁剪，将腋下和上腰部位多余的衣料剪去，并修剪领围、袖口及底摆。

Cut the front and back pieces, the left and right sleeves together, cut off the excess material from the armpits and upper waist, and trim the collar, cuffs and bottom hem.

2. 西式立体裁剪 /Western–style Three-dimensional Cutting

西式裁剪技术不同于传统的直身平面裁剪，作为一种模拟人体穿着状态的分割式裁剪，它可以直接感知成衣的穿着形态及放松量。西式立体裁剪技术的核心主要有斜肩分片装袖、腋下、腰围处收省，同时配以熨烫、归拔技术，使服装更加贴体，更好地展现人体体态美。20世纪30~40年代的改良旗袍就采纳了这种西式裁剪法。

Different from traditional straight flat cutting, the Western–style cutting technology is a split cutting that simulates the wearing state of the human body, which means that it's based on the wearing shape and relaxation of the garment. The core of such technology mainly includes sloping shoulders and separated cut sleeves, with darts at the armpits and waistlines. Meanwhile, ironing and Guiba techniques are used to make the clothing more fitting to better show the beauty of the human body. This cutting technology was used for the improved cheongsam in the 1930s and 1940s.

3. 无省道的裁剪法 /Cutting without Darts

采用收腰方法实现曲线型，其裁剪步骤如下。

The curve shape comes into being by means of waist tightening, and the cutting steps are as follows.

（1）先将面料沿 a 线水平对折，再沿 b 线垂直对折。

First, fold the fabric in half horizontally along line a, and then fold vertically in half along line b.

（2）在折好的面料上画出领口线、侧缝线、小腰、下摆线以及袖口缝线，标出衣长、腰身、挂肩、下摆宽度、下摆起翘量。

Draw the neckline, side seam, small waist, hem line and cuff line on the folded fabric, and mark the length, waist, armholes, hem width, and the jut-out value hem.

（3）在平面空出1个领窝，开1条襟位线。

Leave a collar line on the plane and draw a line for Lapel.

（4）前、后衣片与左、右袖片连在一起裁剪，将腋下和腰身部位的多余衣料剪去，并修剪领围、袖口及底摆。与传统中式平面裁剪方法不同的是，将腰身向内侧挖剪，形成了胸转、腰围、臀围的明显结构，再用归拔的手法对腰身做出强调，使其更加合身。同时还对旗袍的里襟处做了修改，缩小了里襟面积，避免了不必要的重复和堆砌，这种裁剪方法借鉴了西式服装的吸腰与归拔，但尚未涉及省道，所以仍然属于一种部分借鉴的类型。

Cut the front and back pieces together with the left and right sleeve pieces, cut off the excess material on the underarms and waist, and trim the collar, cuffs and bottom hem. Different from the traditional Chinese flat cutting method, the waist is cut inward to make the structure of the chest, waist and hips more prominent, and then use the Guiba technique to highlight the waist to make it fit well. At the same time, the under lap of the cheongsam has been reduced to avoid unnecessary repetition and stacking. This cutting method has referred to the concept of tightening the waist and Guiba of Western clothing, but it has not yet involved the design of darts. Therefore, it remained as a type of partial reference.

4. 有省道的裁剪方法/Cutting with Darts

20世纪40年代的旗袍，衣袖已是单独裁剪，使传统的一块布式的袍身走向了分体组合。通过腋下收省解决了腋下褶皱量的问题，前、后腰身适度的收省量使旗袍更凸显三围的立体感。里襟则继续缩小，并且采用省道合并后裁剪的方法来制作。这种裁剪方法使旗袍更加贴体合身，裁剪步骤为：

For the cheongsam in the 1940s, the sleeves were cut separately, transforming from the traditional one-piece cutting to a split cutting. In such cutting, the amount of folds in the armpit was reduced through the design of darts in this part, and the shape of chest, waist and hip was highlighted by designing appropriate darts in the front and back waist. The underlap was further narrowed, and was made by means of combining darts and cutting. This cutting method makes the cheongsam fit well. The cutting steps include：

（1）先将面料沿 a 线水平对折，再沿 b 线垂直对折。

First, fold the fabric in half horizontally along line a, and then fold vertically in half along line b.

（2）在折好的面料上画出领口线、侧缝线、小腰、下摆线以及袖口缝线，标出衣长、腰身、挂肩、下摆宽度、下摆起翘量。

Draw the neckline, side seam, small waist, hem line and cuff line on the folded fabric, and mark the length, waist, armholes, hem width, and the jut-out value hem.

（3）在平面内空出1个领窝，开1条襟位线。

Leave a collar line on the plane and draw a line for lapel.

（4）前、后衣片与左、右袖片分开裁剪，在胸部、腰围处分别收省，再修剪领围、袖口及底摆。这种裁剪方法采用省道分体式结构，属于全面的借鉴的类型，是近代旗袍接受"西风东渐"历史潮流的代表作。

Cut the front and back parts separately from the left and right sleeves, design darts at the chest and waist respectively, and then trim the collar, cuffs and bottom hem. This cutting method has already adopted the dart split structure, belonging to full reference. It is a representative work of modern cheongsam under the spread of Western influences to the East.

二、旗袍的裁剪步骤/Cutting Steps of the Cheongsam

（一）旗袍的量体方法 / Measuring Method of Cheongsam

量体过程中需注意以下事项：被测量者应穿好紧身内衣；被测量者呈端正、自然的立姿和呼吸状态，不要有多余的动作；在测量"围度"尺寸时，测量者要注意使皮尺保持水平，不宜过松或过紧。

The following matters should be taken into account for the body measurement: the person being measured should wear tight underwear, stand still and breathe naturally with no unnecessary movements; when measuring the "circumference", the tape measure should be kept at an appropriate level that is not too loose or too tight.

（1）总体高（代表服装的"号"）。从头部顶点垂直量至脚跟。

Overall height (representing the "Hao" of the garment). Measured vertically from the vertex of the head to the heel.

（2）衣长。由前身左侧脖根处（肩领点），通过胸部最高点量至所需长度。

Length of clothing. Measured from the root of the neck (shoulder collar point) on the left side of the front body to the required length through the highest point of the chest.

（3）领高。从颈侧点由脖根往上量至所需长度（一般为 3～5cm）。

Neck height. Measured from the side of the neck from the base of the neck up to the required length (usually 3-5cm).

（4）脖颈围。经过第七颈椎点和颈侧点一周的围度。

Neck circumference. The circumference of one circle passing through the seventh cervical vertebra point and the lateral point of the neck.

（5）颈围。颈部最细处水平围量一周。

Neck circumference. The horizontal circumference of the thinnest part of the neck for one circle.

（6）肩宽。手臂自然下垂，左右肩端点之间的长度。注意根据款式要求增加或减少。

Shoulder width. The length between the left and right shoulder end points when the arms hang down naturally, which shall be increased or decreased in line with style requirements.

（7）后背宽。两后外腋点之间的距离。

Back width. The distance between the two posterior outer axillary points.

（8）胸围（代表上衣类服装的"型"）。由前沿腋下，通过胸部最丰满处平衡围量一周。这是紧胸围尺寸，还应按品种要求加放所需松度。

Bust circumference (representing the "shape" of tops). Measured from the axillary point in the front and passes the fullest part of the chest for a circle. This is the tight bust size, and shall be relaxed in line with the clothing types.

（9）胸宽。两前外腋点之间的距离。

Chest width. The distance between the two anterior outer axillary points.

（10）胸高。从颈侧点至乳峰点的距离。

Chest height. The distance from the lateral point of the neck to the point of the breast peak.

（11）乳距。两乳峰之间的距离。

Breast distance. The distance between the two breast peaks.

（12）袖长。从左肩骨外端量至手腕，注意根据不同要求增减长度。

Sleeve length. Measured from the outer end of the left shoulder bone to the wrist, and shall be increased or decreased according to different requirements.

（13）袖口。围量手腕一周，按需要加放松度。

Cuffs. The length of measuring the wrist for a circle, and relax as needed.

（14）腰节。由前身左侧脖根处（肩领点）通过胸部最高处量至腰间最细处。

Waist. Measured from the root of the neck on the left side of the front body (the shoulder collar point) through the highest part of the chest to the thinnest part of the waist.

（15）腰围。在腰间最细处围量一周，并按款式要求放出松度。

Waist circumference. Measured around the thinnest part of the waist for a circle, and shall be relaxed according to the style requirements.

（16）腰长。从腰围线全臀围线之间的距离。

Waist length. The distance from the waist line to the hip line.

（17）臀围。沿臀部最丰满处水平围量一周，并按要求放出松度。

Hip circumference. Measure a horizontal circumference along the fullest part of the buttocks for a circle, and relax the length as required.

（二）旗袍裁剪过程/Cutting Process of Cheongsam

手工制作的旗袍是根据每位顾客的体型进行量体制版，通过样板修正可弥补身材上的不足，使旗袍既合体又美丽。流水线批量生产的旗袍只有几个标准号型的样板，由于不是量体裁衣，所以无法满足各种类型的体型需求。

The hand-made cheongsam is made in line with the body shape of each customer, and can make up for the defect of body shape, allowing the clothing to be well-fitting and beautiful. However, the cheongsam mass-produced by the assembly lines has limited standard models. And such cheongsam cannot meet the needs of various types of body shapes, because it is not tailored.

（1）制作样板。净样板与毛样板都必须核对正确：肩、边、下摆是否等长、等宽，领口、袖笼的弧线长是否与领子、袖山相对应，如图3-9所示。

Make a model. Both the net model and the wool model must be checked correctly to see whether the shoulders, sides, and hem are of the same length and width, and whether the arc length of the neckline and sleeve cap correspond to the collar and sleeve back cap, as shown in Figure 3-9.

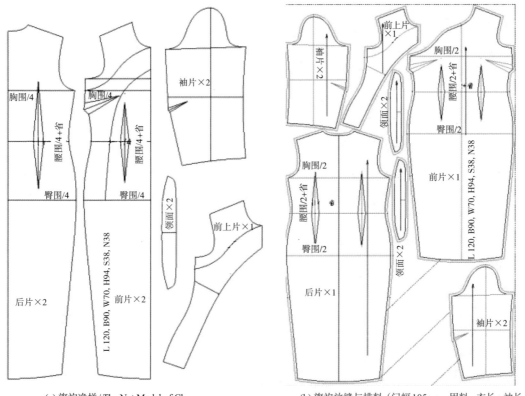

(a) 旗袍净样 / The Net Model of Cheongsam

(b) 旗袍放缝与排料（门幅105cm，用料=衣长+袖长+10cm）/Sewing and Pattern of Cheongsam（Fabric Width 105cm, Material=Clothing Length + Sleeve Length + 10cm）

图3-9　旗袍样板
Cheongsam Template

（2）粉线画样。传统旗袍用料以丝绸为主，丝绸面料质地比较柔软，在拉伸时易变形。在制作时，如果用坚硬的粉饼画线，则会将面料拉斜变形；而用粉线袋，则可避免这个问题。只需按如图3-10所示的方法，轻轻一弹，即刻画出清晰的线条。

Pink line drawing. The traditional cheongsam was mainly made of silk. Because the silk fabric is relatively soft, it is easy to deform when stretched. Therefore, drawing lines with hard power on the silk fabric will make it stretched and deformed. Such a problem can be avoided with the use of pink line drawing. Follow the method shown in Figure 3-10, flick it, and draw a clear line immediately.

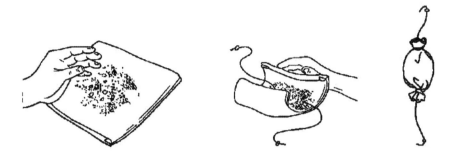

图3-10　粉线袋
Power Thread Bag

（3）裁剪面里料。按照排料图在面、里料的反面进行排料、划样，按照划样线条进行裁剪，注意旗袍的门襟通常在左，里襟在右，正反面有明显区别的面里料排料时不能排成一顺。

Cutting the lining of fabric. Mark and draw the pattern on the reverse side of the surface and the lining in line with the pattern, and cut according to the drawing line. Note that the front placket of the cheongsam is generally on the left, and the inner placket is on the right. The fabric with a distinctive surface and lining shall not be arranged in one line.

三、旗袍制作工具与流程/Cheongsam Production Tools and Process

（一）旗袍制作工具/Cheongsam Production Tools

1. 粉线袋/Pink Power Thread Bag

除丝质面料外，当旗袍用料为棉织物时，也需要使用粉线袋，弹出来的粉线准确而不偏移。传统旗袍虽然造型简洁，但是每一条轮廓线和结构线都精致讲究。粉线袋的使用使传统旗袍在保证其外观造型的前提下，将每一条结构线精练化、准确化。

The pink power thread bag is used for cheongsams made in silk fabrics, as well as cotton fabric. The powder pops out to form straight and accurate lines. Despite the fact that the traditional cheongsam is simple in shape, every outline and structure line is exquisite. The use of the powder thread bag enables the traditional cheongsam to have refined and accurate each structural lines in addition to beautiful appearance.

2. 打水线/Da Shui Xian

打水线是中国传统工艺手法之一，这种工艺手法能够使传统旗袍在造型上更加精致。所谓打水线，就是将棉线含在口中，以口水浸湿，然后将其弹印于刮完糨糊的面料或里料位置，如图3-11所示。打水线的工艺有两个优点：第一，糨糊遇口水软化，极易扣折熨烫，方

图3-11 打水线
Water Line

便定型；第二，由于传统旗袍在造型上讲求简洁，注重线条的流畅感，而打水线的工艺手法恰好易于实现这种造型，所以这种传统的工艺手法一直被延续下来。在旗袍的绳条和衣身的领口、袖口、开衩、底摆等需要扣折熨烫的地方，多采用此方法。

Da Shui Xian is one of the traditional Chinese craftsmanship, which can make the traditional cheongsam more delicate in shape. The technology is to put the cotton thread in the mouth, soak it with saliva, and then print it on the scraped fabric or lining, as shown in Figure 3-11. Such crafting method has two advantages: First, the paste that gets softened by saliva becomes easy to fold and iron for convenient shaping; Second, the traditional cheongsam places great emphasis on simplicity in shape and the smoothness of the lines, and this purpose can be achieved by means of this technology. Therefore, this traditional craftsmanship has been passed on for generations, often used in places that need to be folded and ironed, such as the neckline, cuffs, slits, and bottom hem of the cheongsam.

3. 刮糨糊/Scraping Paste

传统绳条的制作工艺必须满足两个条件。首先绳边布必须刮糨糊，其次要以45°的正斜纱裁剪绳条。按照图3-12所示的方法刮糨糊，用刮刀取适量糨糊，在绳边布的反面沿着直纱方向刮一至两遍，待晾干后直接用木尺在绳边布的反面沿着横纱方向反复刮拭，使其由坚硬变得柔软。最后，用干熨斗在其反面干烫定型即可。这种工艺手法可以固定住横纱和直纱的丝道方向，在绱绳条的时候不易撑拉变形，出现绳边宽窄不均匀的情况而影响整体造型上的美观性。另外，刮糨糊的工艺手法可以与打水线结合使用，使绳条易于扣折熨烫成所需要的宽度，并且保证宽窄的一致性。无论是0.2cm的线香绳，还是宽绳边，都可以在工艺手法上实现。

Two conditions have to be met for the production process of traditional rolling welts. The piping cloth must be scraped with paste and cut into welts with a 45°positive bias yarn. Scrape the paste, as shown in Figure 3-12, use a scraper to take an appropriate amount of paste, and scrape it once or two times on the reverse side of the piping cloth along the straight yarn direction. After drying, use a wooden ruler directly on the reverse side of the piping cloth repeatedly along with the horizontal direction of the thread to make it soft, and finally, dry and iron the reverse side to set the shape. This process can fix the thread direction of the horizontal yarn and the straight yarn, and it is not easy to stretch and deform when rolling the scraper, while the uneven width of the piping will affect the aesthetics of the overall shape. In addition, the

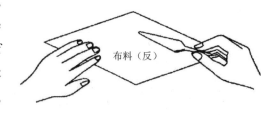

图3-12 刮糨糊
Scraping Paste

process of scraping paste can be used in combination with the Da Shui Xian, so that the roller can be easily folded and ironed to the required width, and ensure the uniformity of width. Both the 0.2cm incense roll and the wide piping can be achieved with this technique.

4. 斜裁/Bias Cut

斜裁是以45°正斜纱的方向斜裁成绲条，如图3-13所示。这种做法可以保证在绲绲条时，绲条布可以适合曲线边缘弯成多变的形状。另外，在后期手针撬缝绲条反面时，缝纫线以45°倾斜角嵌入丝道中不显露在外，保证了细节上的美观性。

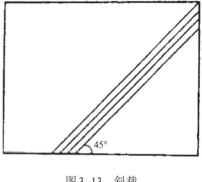

The method of bias cut is to diagonally cut the fabric into rolling strips in the direction of 45° positive bias yarn, as shown in Figure 3–13. This method can ensure that when rolling, the rolling cloth will be bent into a variable shape in line with the curved edge. Additionally, when the hand needle sews the reverse side of the roller, the sewing thread is embedded in the thread at an inclination angle of 45° and will not be exposed, to achieve beautiful details.

图3-13 斜裁
Bias Cut

（二）旗袍制作流程/Cheongsam Production Process

（1）收省道。先沿粉线记号烫出中心线，疏缝固定。照记号车缝，省尖要用车缝线打结，不可回针。省道左右分开烫平，省道小、省份少时，倒向一边烫平。

Dart design. First，iron the center line along the pink line mark, and fix it with sparse seams. Sew according to the mark, and the sharp point of the dart should be knotted with sewing thread with no backstitch. Iron both sides of the dart in sideways. The smaller the dart is, the less time it takes.

（2）底襟处理。折0.7~1cm，车缝0.8cm宽。

Bottom placket. Fold 0.7–1cm, and sew 0.8cm wide.

（3）大襟的处理。先缝1cm宽牵条，再裁剪贴边。

Processing of the big placket. First，sew a 1cm–wide drawstring, and then cut the clipping edge.

（4）后身缩烫并缝牵条。

The back is shrunk and ironed, and sewn with the drawstring.

（5）前身缩烫并缝牵条。

The front is shrunk and ironed, and sewn with the drawstring.

（6）开衩处垫叉布。

Add the fork strip to the openings.

（7）合肩线。车缝时领口处再回针4~5cm，避免剪领口后分开。

Shoulder closing line. Back stitch about 4–5cm at the neckline when sewing to avoid separation after cutting the neckline.

（8）裁剪袖口贴边。

Cut the cuff welt.

（9）做领子。

Make the collar.

（10）开领口。用皮尺立起来量，因包括缝份，故比实际的领围少3~4cm。

Open the neckline. Use a tape measure to measure it upright. Since it includes seam allowances, it is 3–4cm less than the actual neckline.

（11）绱领子。回针缝固定，必须将领衬藏住，但不缝到表布。

Upper collar. Backstitch to fix to hide the collar lining without sewing the table cloth.

（12）接合边缝。

Join the side seam.

（13）画盘扣位置及钉暗扣记号，底襟暗扣下垫布，缝领勾。

Draw the position of the knot button and the mark of the hidden button, add the bottom placket with the hidden button and the underlayment cloth, and sew the collar hook.

（14）上拉链。齿入0.1cm，用星点回针缝固定。

Upper zipper. The teeth are inserted into 0.1cm and fixed with a star point back stitch.

（15）缝暗扣。领上0.8cm先缝领钩，领角缝暗扣。

Sew the hidden button. First, sew the collar hook 0.8cm up on the collar, and sew the hidden button on the collar corner.

（16）下摆与边开衩。固定下摆使用斜针或藏针缝，边用千鸟缝。

Hem and side slits. Use oblique stitches or hidden stitches to fix the hem, and sew the sides with the Qianniao sewing method.

（17）开衩止点、拉链止点固定防裂开。

The stop point of the slits and the stop point of the zipper are fixed to prevent cracking.

（18）整烫。

Ironing.

第二节　旗袍制作工艺/Cheongsam Production Process

一、前期准备/Preliminary Preparation

传统旗袍和现代旗袍在制作工艺上具有不同的特点。传统旗袍在制作工艺特点是做工精良考究，多采用手工刺绣、镶、嵌、绲边等工艺技法；现代旗袍工艺特点是式样简洁合体的线条结构，代替了精细的手工制作。

Traditional cheongsam and modern cheongsam have different characteristics in the production process. The former is characterized by its fine and exquisite workmanship, with many techniques such as hand-embroidery, inlay, and sewing on the edges used. The latter is characterized by its simple and integrated line structure, replacing fine handwade craftsmanship.

（一）核对裁剪衣片及制作标记/Check the Cut Pieces and Make Marks

在缝制前，首先要检查旗袍的裁片，依次核对面料、里料及辅料的质量和数量，并放整齐。然后根据旗袍的材质特性及部位选用线丁、粉印、眼刀、针眼等不同的方法做出标记。

Before sewing, the cut pieces of the cheongsam shall be first checked to see if the quality and quantity of the fabrics, linings and accessories are correct, and placed neatly. Then based on the material characteristics and parts of the cheongsam, different methods such as thread, powder printing, and eye knife and needle eye are used to make marks.

（二）制作省道/Making Darts

按缝制标记缉省，尽量与人体体型相吻合。一般材料的省缝制作可倒烫，在熨烫省道的同时将胸部胖势烫出，通过归拔技术将腰节部位拔开，使省缝平服且不起吊；精良的面料在缝制省缝时采用中间分烫的方法，表面平服无折痕，注意熨烫时要根据面料特性选择不同的温度，或干烫，或湿烫。

Darts are made according to the sewing mark to make the clothing perfectly fitted with the shape of the human body. The dart seam production of general materials can be reverse ironed. While ironing the darts, the chest curve shall be ironed in the meantime, and the waist part is pulled out through the Guiba technology, so that the dart seam is not only flat but also stays still; the method of ironing in the middle is used when saving dart seams on de superior fabric to make the surface flat without creases. Note that different temperatures shall be selected to conform to the fabric's features, using ironing or wet ironing.

（三）衣身归拔/Guiba Technique

旗袍造型在线条上要求流畅，人体穿着要合体无褶皱，制作中只有通过传统工艺中的归拔技巧才能达到效果。需要注意的是，不同材料的耐热度不同，熨烫时要选用适当的温度进行归拔。前衣片需要归拔胸部、腹部、摆缝及肩缝，后衣片需要归拔袖窿、背部、臀部、肩缝及摆缝，通过归拔工艺使衣片更加贴合人体体型特征。

The Guiba technique refers to the craftsmanship that stretches the garment fabric to appropriately change its warp and weft structure, to lengthen（pull）, or shorten（return）, or push in one direction（push）, so as to get the three-dimensional shape of the garment. Only by such technology can the shape of the cheongsam be smooth in lines and a perfect fit without wrinkles. It should be noted that different materials have different heat resistance, and the appropriate temperature should be selected when ironing. The chest, abdomen, hem seam and shoulder seam in the front piece need to be pulled and pushed, as well as the armholes, back, buttocks, shoulder seam and hem seam in the back piece. Such technology enables the garment to better fit the shape of the human body.

（四）黏附牵条 /Adhesive Strip Cloth

为避免归拔后衣身制作中二次变形，需要黏附牵条。要求将牵条粘贴在净粉线上，部位包括前襟斜襟、前后片袖窿、摆缝到开衩的位置，注意在臀部附近处将直丝牵条粘得略紧一些。

In order to avoid secondary deformation in the production of the clothing after Guiba technology, it is necessary to paste a strip cloth. It is required to stick the strip cloth on the net powder line, including the front slanted placket, the front and rear armholes, the position of the hem to the slit, and the straight yarn cloth shall be pasted a little tighter near the buttocks.

（五）绲边 /Piping

根据造型工艺的要求不同，绲边的制作方法有多种多样。传统的手工工艺采用暗线绲边的方法，首先将用熨斗折净衣片毛缝，开衩处剪一眼刀至净粉0.1cm处，将绲条缉缝宽度为0.4～0.5cm，然后将绲条翻转、翻足，再将绲条包转、包足，接着将绲条反面与大身撬牢，注意不能撬到衣身的正面，最后将夹里盖过撬线与绲条撬牢，此工艺的特点是绲边饱满、完整，要求在衣身的正面不露出针迹。

According to different requirements of the molding process, methods for making piping vary in a wide range. The traditional handicraft adopted the method of sewing the piping with dark thread. First, fold the woolen seam of the net piece of clothing with an iron, cut at the opening to 0.1cm away from the net powder, and sew the stitching width to 0.4–0.5cm. Turn the piping over for enough times, then wrap it and inlay the back of the piping to the clothing body, be careful not to pry the front of the clothing, and finally cover the inlaying thread with lining and make it stick to the piping to make piping fully covered and complete. No stitches shall be exposed on the front of the clothing.

二、款式制作 /Style Production

（一）缝合肩缝并缉袖子 /Sew the Shoulder Seam and the Sleeves

首先将前、后衣片的正面相对并对齐肩缝，前片在上面进行缉缝小肩缝，注意后肩缝要略有吃势以符合人体的需求，然后将肩缝进行劈烫。制作袖口的绲边方法与摆缝开衩方法相同，调整好袖山的吃势与袖窿相对进行缝合。

First, align the front sides of the front and back parts with the shoulder seam, and sew the small shoulder seam on the front part. Note that there should be ease for the back shoulder seam to meet the requirement of the human body, and then split and iron the shoulder seam. The method of making the edge of the cuff is the same as that of the seam and the opening, and then adjust the ease of the sleeve cap and sew the armhole.

（二）做夹里 /Make a Folder

将夹里省缉好，缝合肩缝夹里，再装袖夹里，缝好后将省缝、小肩缝进行倒烫，要求熨烫平服，里子底边比面料短1cm，折净后固定。

Make a dart and sew it into the shoulder seam, and then install it in the armhole. After sewing, reverse the dart seam and the small shoulder seam, and get them ironed to be flat, and the bottom edge of the lining is 1cm shorter than the fabric. Fold and fix it.

（三）勾缝夹里/Sew the Linings

夹里与小襟正面相对并对齐里面的领口、肩缝，然后沿净粉线绲缝并翻烫平服。大襟及开衩部位的毛边折净，要求压过绲边的绲缝线，然后选用手工撬缝工艺撬牢固。后片夹里的制作与前片夹里的勾缝方法相同。

Align the lining with the front of the small placket and align the neckline and shoulder seams inside, then stitch along the clean pink line and turn over and iron the garment. The raw edges of the lapel and the opening should be folded well to press the stitching line of the piping, and then the manual prying process is used to pry it firmly. The production of the lining if the back is the same as that in the front.

（四）缝合摆缝、袖缝/Sew the Hem Seam and the Sleeve Seam

先将前后衣片的正面相对，对准各部位后沿净缝绲线，然后分烫摆缝、袖缝的缝边，翻正衣片后再将袖口夹里折光，盖过绲边绲线，然后撬牢固。

First, make the front and back sides of the front and back pieces face each other, align each part and follow the clean seam line, then iron the hem seam and the sleeve seam, turn over the garment pieces, and then well fold the lining of the cuff to cover the seam line and pry firmly.

（五）做领、绱领/Make a Collar and Connect It with the Clothing

将净领衬烫在领面的反面，领面上口沿衬边缘包转，将绳条绲缝在领面上口，领面与领口正面相对沿领衬下缘进行绲缝。绱好的领子需要检查领面是否圆顺、平服，领子左右是否对称，各对位点是否准确圆顺。然后扣净领里的缝头并与领面反面撬牢，注意领里略紧于领面，如图3-14所示。

Iron the net collar lining on the reverse side of the collar, wrap the opening of the collar along the edge of the lining, sew the tape on the upper edge of the collar, and sew along the lower edge of the collar lining opposite the front of the collar. For a well-twisted collar, it is necessary to check whether the collar is round and flat and symmetrical from left to right, and whether each pair of points is accurate and round. Then clear the seam in the collar and pry it with the reverse side of the collar. Note that the collar lining is slightly

图3-14　绱领
Connect Collar

tighter than the collar, as shown in Figure 3–14.

三、零部件制作/Parts Production

（一）制作、钉缝盘扣/Making and Sewing Knot Buttons

裁剪2cm左右的斜条布，将两边毛口向里折，然后对折层，选用手针进行撬缝牢固，如图3-15所示。如果选用的是较薄的材料，可以将斜料裁剪得宽一些，多折几层再进行撬缝，也可在斜条中加几根纱线，这样制作的盘扣条显得比较饱满。为了便于盘花造型保形，撬钮袢时经常加入细铜丝，盘花是将盘扣条盘结成所需的各种形状并用线钉好，盘缝花扣需要注意条、花的比例协调，大小规格可根据款式和花型确定。

Cut a diagonal cloth of about 2cm, fold the burrs on both sides inward and then fold the layers in half, and use a hand needle to pry the seam firmly, as shown in Figure 3–15. When it comes to a thinner material, the diagonal material can be cut wider and folded with a few more layers before prying, or a few yarns can be added to the diagonal strip. The knot button strip made in this method looks plump. To maintain the shape of the pattern of the button, fine copper wire is often added when prying the button loop. The pattern of the button is to form the buckle strip into various shapes required and nail it with thread. In this process, it should be noted that the ratio of strips and flowers, and the size specifications can be determined according to the style and the shape of the pattern.

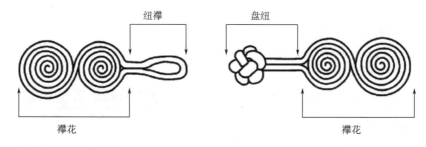

图 3-15 盘扣
Knot Buttons

（二）钉领钩，打套结/Nail Collar Hook, and Make Tie Knot

领钩钉缝在大襟一侧领的圆角处，领袢钉缝在小襟一侧领的圆角处，要求均与领止口平齐，左右高低一致，钉缝牢固。为了增加牢固程度，可在圆角处钉上按扣，不要露出针脚。衣身两侧的摆缝开衩处采用套结针法，用手针将衩口封牢固。

The collar hook is stitched to the rounded corner of the collar on the side of the large placket, and the collar tab is stitched to the rounded corner of the collar on the side of the small placket. It is required to be flush with the neckline, and the left and right heights are consistent and the stitching is firm. Snaps can be attached to the rounded corners for security purposes, but no stitches are exposed. The hem slits on both sides of the clothing are securely closed with the bar–tack stitch.

（三）整烫/Ironing

旗袍制作完成后，先修剪线头及检查并清洗污渍，通过整烫使服装平服并符合人休休形特征。整烫顺序：先烫里后烫面，先烫附件后烫主件，从上至下进行各部位的熨烫。

When the cheongsam is complete, trim the thread ends and check and clean the stains on it, making it flat to fit the shape of human body by ironing. Ironing sequence：iron the inside first, then the surface；iron the accessories first, then the main piece, and iron each part from top to bottom.

思考题

（1）旗袍制作包括哪些步骤？

（2）简述旗袍的成型过程。

Questions

（1）What are the steps involved in making a cheongsam?

（2）Briefly describe the forming process of cheongsam.

○ 第四章

旗袍材质之美
The Beauty of Cheongsam Materials

第一节　旗袍与丝绸面料/Cheongsam and Silk Fabrics

　　旗袍作为我国的传统服饰，以其独特的款式造型、精致的剪裁和制作工艺，一直深受消费者的喜爱。旗袍通常采用丝绸面料、绒料及棉、麻、毛呢绒面料等制作，应用较多的就是丝绸面料，丝绸面料种类繁多，其特点不尽相同。

As a traditional costume of China, cheongsam has always won consumers' favorite for its unique style, exquisite tailoring, and refined production technology. The fabrics used to design and make cheongsam usually include silk, fleece materials and cotton, linen, woolen. The most widely used is silk fabrics which are wide-ranging. Different silk fabrics have different characteristics.

一、旗袍面料分类与特性/Classification and Characteristics of Cheongsam Fabrics

（一）真丝类/Real Silk

　　真丝是旗袍的主要制作材料之一，它具有良好的吸湿性和透气性，触感柔软细腻，光泽柔和。真丝顺滑的手感和良好的垂坠感使丝质旗袍充满了奢华贵气，很适合在正式场合穿着。旗袍面料大多采用真丝绸缎，常用于制作旗袍的丝绸包括织锦缎、丝绒、乔其纱等，尤其适合夏季穿着。

Real silk is one of the main production materials of cheongsam that is soft and delicate, featuring prominently in good moisture absorption, good breathability, graceful luster, smooth feeling, and good drape. The silk cheongsam looks very luxurious and noble, suitable for formal occasions. Most of the cheongsam fabrics are made of silk satin, and the commonly used silk includes brocade, velvet, georgette, etc., which are mostly worn in summer.

　　丝绸面料的加工方法不同、经纬原料不同、织物组织不同及颜色搭配的不同，面料的风格也是千变万化，品种繁多。在旗袍设计中，常用的真丝面料有以下几类：

Because of its different processing methods, warp and weft raw materials, fabric structures and colors, the styles of silk fabrics are ever-changing with diverse varieties and rich and colorful knowledge content. The silk fabrics commonly used in cheongsam design are as follows.

1. 真丝缎类面料/Real Silk Satin

缎类面料大部分采用缎纹组织织成的花、素织物。织纹造型细秀工整，色彩丰富，质地柔软，平滑光亮。其中，最有代表性的是花软缎、金玉缎、古香缎、花蕾缎和花库缎等。缎类面料是工艺上最为高级的一个大类，组织结构千变万化，外观绚丽多彩。在旗袍设计中常用的缎类面料有素绉缎、弹力素绉缎、素缎、暗花缎、妆花缎等。暗花缎是指正反缎纹，正面是经面缎纹，起花部分是纬面缎纹；或者地组织是纬面缎纹，起花部分是经面缎纹。

Most of the satin fabrics are flower and plain fabrics woven with satin weave. The texture is fine and neat, rich in color, soft in texture, smooth in feeling, and bright in luster. The most representative ones consist of mixed satin brocade, Jinyu satin, Suzhou brocade, bud satin brocade, palace satin and so on. Satin fabrics, the most advanced category in terms of craftsmanship, boast ever-changing structures and the most colorful appearance. The commonly used fabrics in the design of cheongsam include plain crepe satin, stretch plain crepe satin, plain satin, dark satin, Zhuanghua satin, etc. Dark satin refers to the satin that has warp satin weave in the front and back, and weft satin weave at the embossed part, or has weft satin weave in the ground and the warp satin at the embossed part.

（1）素绉缎。素绉缎是丝绸面料中的常规面料，亮丽的缎面显得非常华贵，手感滑爽，组织密实；但是，这种面料的缩水率相对较大，下水后光泽有所下降。

Plain crepe satin. Plain crepe satin is a common fabric in silk fabrics with a bright and noble satin surface, smooth feeling, and dense weaving. However, the shrinkage rate of the fabric is relatively large, and the luster decreases after watering.

素绉缎拥有如同珍珠般的顺滑光泽，色彩亮丽，由于这种面料容易起皱，所以经过熨烫平顺后才能完美地展现它的光泽；运用这种面料制作的旗袍外观华美、高贵而极具高档感，如图4-1所示。

Plain crepe satin feels smooth, with shiny luster and bright color. Because this fabric is easy to wrinkle, its luster can only get fully demonstrated after ironing. The cheongsam made with satin looks very gorgeous, noble and high-end, as shown in Figure 4-1.

图4-1 素绉缎面料
Plain Crepe Satin Fabric

（2）弹力素绉缎。弹力素绉缎是新面料，成分为90%～95%桑蚕丝，5%～10%氨纶，属于交织面料。其特点是弹性好、舒适，缩水率相对较小，风格独特。这种面料不是100%的真丝，加入了氨纶，面料有弹性，悬垂性好。真丝比重越大，手感越接近绸缎，光泽也越漂亮。

Stretch crepe satin. Stretch crepe satin is a new, interwoven fabric, with 90% to 95% composition of mulberry silk and 5% to 10% spandex. It is characterized by good elasticity, comfort, small shrinkage rate and unique style. This is not 100% silk, added with other ingredients, having good elasticity and drape. The greater the proportion of silk, the more it feels like satin with brighter luster.

2. 香云纱面料/Gambiered Canton Gauze

香云纱是原生态传统面料，以桑蚕丝为原料，属于比较高端的面料。用香云纱做成的旗袍，穿着者走动时会产生沙沙声，还具有防水防晒的优点。香云纱面料色泽古朴怀旧，经久耐穿，手感挺爽柔润，具有古典韵味。

Gambiered canton gauze is an original ecological traditional fabric, with mulberry silk as the raw material, belonging to a relatively high-end variety. The cheongsam made of such fabric rustles when walking, and is waterproof and sunscreen. Its color of gambiered canton gauze is simple and nostalgic, its material is durable, and the feeling is cool and soft, with a classical charm.

香云纱是一种古老的手工织造和染整制作的植物染色面料，已有一百多年的历史。它制作工艺独特，数量稀少，制作时间长，技艺要求精湛，穿着时滑爽、凉快，并具备除菌、驱虫、保健皮肤的功效。因穿着后涂层慢慢脱落，露出褐黄色的底色，它被形象地称为"软黄金"。它也是常用的一种以真丝为主的面料。

This gambiered canton gauze is an ancient plant-dyed fabric made by hand weaving and dyeing and finishing. It has a history of more than 100 years. Benefiting from its unique production process, rare quantity, long production time, and exquisite techniques, it feels smooth and cool, has the function of sterilization, repelling insects, and is healthy for the skin. When the color fades away gradually after wearing, the brownish-yellow bottom color gets revealed, thus being called the soft gold vividly in the past time. It is also a commonly used silk-based fabric.

香云纱是用广东特有的植物"薯莨"染制而成，前后经过30多道工序，染织晾晒需要60多天，有一种与生俱来的大自然植物气息。香云纱色泽古朴怀旧，衣物有易干的特性，抗皱性和还原性都较普通真丝好，属国家非物质文化遗产，如图4-2所示。

The source of gambiered canton gauze is the dyed and woven fabric from the unique plant of Guangdong, "Dioscorea cirrhosa Lour." More than 30 processes are involved before and after, and it takes more than 60 days for dyeing, weaving, and drying. It has an innate natural plant smell, and the color is simple and nostalgic. The clothing made in such fabric is easy to dry, with better performance in wrinkle resistance and reducibility than ordinary silk, listed as the national intangible cultural heritage, as shown in Figure 4-2.

3. 织锦类面料/Brocade Fabrics

在旗袍设计的面料选择中，织锦类面料因其工艺精湛，图案华美，常用于高端旗袍设

图4-2　香云纱面料旗袍
Gambiered Canton Gauze Cheongsam

计，中国织锦包括四大名锦，即云锦、蜀锦、壮锦、宋锦。

Among the fabrics for cheongsam, brocade fabrics are often used in high-end cheongsam designs because of their exquisite craftsmanship and gorgeous patterns. There are four famous Chinese brocades, namely Yun Brocade, Shu Brocade, Song Brocade and Zhuang Brocade.

（1）云锦。云锦源自南京，可追溯至宋朝，流行于明清时期，曾是皇家御用贡品，被誉为中国古代织锦工艺史上的里程碑。云锦距今已有1580多年的历史，因其用料考究、织造精湛、图案色泽绚烂若云彩而得名。云锦的传统品种有妆花、织金、库缎、织锦等。将长跑梭、短跑梭及吸收缂丝通经断纬技法的挖花梭，结合运用于提花丝织物的妆彩上，采用缎、纱、罗、绸、绒各种质地的丝织物。云锦的花纹配色极度自由，色彩变化多样，其织造技艺已被列入国家非物质文化遗产。由于云锦使用长跑梭、短跑梭及挖花梭，在纬向上有不同颜色，并且颜色也不受限制，因此只能使用手工织造，即使用计算机控制的现代织机也无法织造云锦。

Yun Brocade. Yun Brocade originated in Nanjing and can be traced back to the Song Dynasty. It was popular in the Ming and Qing Dynasties and was a tribute to the royal family, was called the milestone in the history of ancient Chinese brocade craftsmanship. With a history of more than 1, 580 years, the brocade has been famous for its exquisite materials, refined weaving, gorgeous patterns, and cloud-like color. Its traditional varieties include Zhuanghua (the silk fabric with patterns woven in colorful thread), Zhijin (the silk fabric with patterns woven in gold thread), palace satin, brocade and so on. It applied the Changpao shuttle (color shuttle for the uniform color), Duanpao shuttle (color shuttle for segmented color shift), and Wahua shuttle (the weaving techniques that refers to the silk tapestry, with cut designs) into the color of jacquard silk fabrics, using silk fabrics of various textures such as satin, gauze, Luo, silk, and velvet. The pattern and color matching of Yun Brocade can be matched freely, the colors vary from a wide range, and its weaving skills are included in the national intangible cultural heritage. The use of the Changpao shuttle, Duanpao shuttle, and Wahua shuttle, the brocade has different, unlimited colors in the weft. It can only be woven by hand, even the modern computer-controlled looms cannot make it.

（2）蜀锦。蜀锦产自四川成都一带，因其历史悠久、工艺精湛、图案华美，成都因此得名"锦官城"。蜀锦是丝绸之路上的主要交易品之一。蜀锦大多以经线彩色起彩，彩条添花，经纬起花，先彩条后锦群，方形、条形、几何骨架添花，对称纹样，四方连续，色调鲜艳，对比强烈，是一种具有汉民族特色和地方风格的多彩织锦。蜀锦的织造工艺有独特的整经手法，至今蜀锦仍沿用传统的染色熟丝织造法。2006年蜀锦织造技艺被列入国家级非物质文化遗产名录。蜀锦图案对称，擅长表现动物、植物、字纹、器物及风景。由于蜀锦是通经通纬织造，因此在现代织机上也可以织造。

Shu Brocade. Shu Brocade was produced in Chengdu, Sichuan, its long history, exquisite craftsmanship and gorgeous patterns allowed Chengdu to be named the "Jin Guan City". Shu Brocade almost became one of the main trade items on the Silk Road. Most of the brocades were colored with warp threads, decorated with colored strips, square or strip shapes, symmetrical patterns and bright colors. It is a kind of colorful brocade with the characteristics of Han nationality and the local style of Chengdu. The weaving process of Shu Brocade adopts the unique warping technique. So far, Shu Brocade still follows the traditional dyed and cooked silk weaving method. In 2006, Shu Brocade weaving skills were included in the national intangible cultural heritage lists. The patterns on the brocade are symmetrical, mainly including animals, plants, characters, utensils. Because Shu Brocade is woven through warp and weft, it can be woven on modern looms.

（3）壮锦。壮锦又称僮锦，是广西壮族自治区的著名丝织物，以棉线或丝线编织而成，图案生动，结构严谨，色彩斑斓，充满热烈、开朗的民族特色。

Zhuang Brocade. Zhuang Brocade, also known as Tong Brocade, is a famous silk fabric in Guangxi Zhuang Autonomous Region. It is woven with cotton or silk threads, with vivid patterns, rigorous structure, and diverse colors, presenting a warm and cheerful national style.

（4）宋锦。宋锦源自苏州，始于宋末，在唐代蜀锦的基础上发展而来。宋锦色泽华丽，图案精致、质地坚柔，被誉为中国"锦绣之冠"，已被列入"世界非物质文化遗产"。它将"针尖上的文化"传递给世界。宋锦是我国的四大名锦之一，至今仍焕发独特魅力。它是在特定的历史背景下，在历史、政治、经济的共同影响下，而造就的具有富丽典雅美学特色的织锦。宋锦织造工序繁多复杂，制作精细讲究，彰显深厚的文化底蕴。宋锦形成于两宋，在当时的社会背景下，带有宋代上层社会的审美特征。宋锦是指以经纬线同时显花的织造技艺，在继承宋代织造艺术特色的基础上，经过元、明、清三朝的发展，形成了一种以经面斜纹作地、纬面斜纹显花的特色锦，又称为宋式锦或仿宋锦，统称为宋锦。苏州得益于地域优势，经济富饶，因此丝织技术发达，素有"丝绸之府"之称。由于当时苏州地区的织造技术水平较高，所生产的宋锦质量上乘，因此宋锦逐步兴起，又称为"苏州宋锦"。

Song Brocade. Song Brocade originated in Suzhou, began in the late Song Dynasty, and developed based on Shu Brocade in the Tang Dynasty, with gorgeous color, delicate pattern, and soft texture, and it is praised as China's "Best Brocade" and is included in the World Intangible Cultural Heritage, spreading Chinese weaving culture to the world. Moreover, Song Brocade is one of the four famous brocades in China. This splendid and graceful brocade came into being against the specific historical background and under the joint influence of history, politics and economy. And its weaving techniques well presented the profound culture heritage of Chinese nation with numerous and complicated processes and meticulous production methods. The aesthetic characteristics of the upper-class society of the Song Dynasty can be seen in this brocade which was shaped during this period in the context of the social environment of then. Song Brocade refers to the weaving technique that makes patterns with both the warp and weft threads. It developed on the basis of the weaving arts of the Song Dynasty, and formed the unique brocade with the warp to make the ground and the weft to make the patterns after the development throughout the three dynasties of Yuan, Ming, and Qing, also known as Song-style brocade and Song-imitated brocade. Benefiting from its geographical advantages, Suzhou has a prosperous economy and developed silk-weaving technology. The superior regional advantage contributed greatly to the economic growth and advanced weaving technology of Suzhou, allowing it to be known as the "Home to the Silk". And owing to the high level of weaving technology of Suzhou at that time, the Song Brocade produced here was of high quality. During this period, Song Brocade became increasingly popular, so it was also called the "Suzhou Song Brocade".

宋锦品质精美，价格高昂，与旗袍的品质风格相匹配，尤其适合用于高端旗袍设计中，其质地柔软，花纹精致，色泽高贵，美轮美奂，与旗袍精致的工艺相结合，能够很好地展现旗袍的神韵，如图4-3所示。

Song Brocade is exquisite in quality, and expensive in price, which matches the quality and style of cheongsam, especially suitable for high-end cheongsam design. Its soft texture, delicate patterns, graceful color, and excellent luster can express the charm of cheongsam to the fullest with the refined production process of cheongsam, as shown in Figure 4-3.

4. 丝绸罗类面料/Leno Silk Fabric

丝绸罗类面料是中国丝绸代表品种"绫罗绸缎"中的一类。罗类面料是一种比较轻薄通孔的丝织物，其外观稀疏、有孔隙和绉感。通过绞纱和平纹的交替织造，经丝互相纠结，形成条状孔路，这些孔眼疏朗且稳定。罗类面料可分横罗、直罗和花罗。杭罗已被列入国家非物质文化遗产名录，并与苏缎、云锦一同被列为中国东南地区的三大丝绸名产。杭罗原产杭州，故名杭罗，以纯桑蚕丝为原料，通过平纹和纱罗组织联合构成，绸面具有等距规律的直条纹或横条纹菱形纱孔，孔眼清晰，质地刚柔滑爽，穿着舒适凉快，耐穿，耐洗。杭罗尤其适合在闷热多蚊虫天气穿着，既挺阔、透气，又可防止蚊虫叮咬，这也是杭罗在古代作为宫廷御用衬衣面料的原因。现代旗袍设计将罗类面料与里料相结合，具有通透、层叠的艺术美

图4-3 宋锦面料旗袍
Song Brocade Fabric Cheongsam

感，通常应用在夏季旗袍的设计中，如图4-4所示。

Leno silk fabric is one of the most representative varieties of Chinese silk that is light and thin with through-holes, characterized by sparseness, interspace, and a feeling of crepe in appearance. It is woven by the alternation of twisted skein and plain weave, and the warp threads are entangled with each other to present strip-shaped holes that are sparse and stable. This fabric can be divided into horizontal leno, straight leno, and flower leno. The famous Hangzhou leno is included in the National Intangible Cultural Heritage list, and it is listed as the three famous silk products in southeastern China together with Suzhou satin and Yun Brocade. Hangzhou leno originated in Hangzhou, with pure mulberry silk as raw material and composed of plain weave and leno weave, having equidistant and regular straight or horizontal stripes with diamond-shaped, clear gauze holes. It feels soft, smooth, and cool, and it is durable and washable, very suitable for hot and mosquito-rich weather. Its excellent shape and good breathability, coupled with mosquito prevention made Hangzhou leno the fabric used for royal clothing in ancient times. Leno fabric and lining are combined for the design of modern cheongsam, to achieve unique patterns that are transparent with artistic beauty, which is usually used in the design of summer cheongsam, as shown in Figure 4-4.

图4-4 罗面料
Leno Fabric

5. 绸类织物/Plain Woven Silk Fabric

绸类织物的地经采用平纹或各种变化组织，或同时混用几种基本组织和变化组织。无论用单梭还是多梭，无论单经轴还是多经轴，织造的花、素织物，只要没有其他特征，均可归入绸类。例如塔夫绸，它的经丝采用复捻熟丝，纬线采用并合单捻熟丝，以平纹组织为地，织品密度大，是绸类织品最紧密的一个品种之一。真丝塔夫绸柔而平整，薄而丰满，其组织结构是细腻的平纹和缎纹。塔夫绸有素塔夫、条纹塔夫、花塔夫之分。塔夫绸的保形性极佳，有丝鸣现象，尤其适合制作婚纱、晚礼服、旗袍等。

The ground warp of the plain woven silk fabric refers to the use of plain weave or various changed structures, or a mixture of several basic structures and changed structures at the same time. Regardless of single–shuttle or multi–shuttle, and flower and plain fabrics woven with single and multi–warp beams. Fabrics with no other characteristics can be included in this category. Such as taffeta, its warp is made of double–twisted cooked silk, and its weft is made of combined single–twisted cooked silk. The plain weave is used as the ground with high density, thereby regarded as one of the most compact variety of silk fabrics. The silk taffeta is characterized as soft and flat, thin and plump. The weave is fine plain and satin. Taffeta includes plain taffeta, striped taffeta, and pattern taffeta. Such fabric features prominently in keeping in shape, especially suitable for wedding dresses, evening dresses, cheongsams, etc.

（1）双宫绸。双宫绸一般是用桑蚕双宫茧为原料缫制的丝织造，也可在双宫茧批中掺入一定比例的上茧或次茧混合缫制。双宫绸的特征是绸面呈现明显的不规则疙瘩，质地坚挺厚实，织物别具风格，因为双宫绸表面存在闪光和疙瘩，因此也称疙瘩绸。双宫绸质地挺括，有光泽，但它的光泽不像缎类面料那么光亮，而是比较内敛低调。双宫绸价格比较昂贵，属于高档真丝面料。表面呈现的疙瘩使制成的旗袍更有一种肌理风格，如图4-5所示。

Doupion silk. Doupion silk is usually made of the silk reeled from silkworm cocoons, and it can also be mixed with a certain proportion of first–class cocoons or secondary class cocoons in the batch of double cocoons. Dupion silk is characterized by obvious irregular bumps on the surface, firm and thick texture, and a unique fabric style. Because of the special style of glitter and bumps on the surface, it is also called Geda Silk（silks with lumps）. Its texture is crisp and shiny, with a reserved luster that is not as bright as satin, but looks very high–end. Therefore, this kind of silk has been included in the top–end silk fabrics which are expensive. The lumps on the surface of the fabric make the cheongsam made in this

图4-5　双宫绸面料
Doupion Silk Fabric

have a textured style, as shown in Figure 4–5.

（2）塔夫绸。塔夫绸是一种以平纹组织织制的熟织高档丝织品，经纱采用复捻熟丝，丝采用并合单捻熟丝，以平纹组织为地，织品密度大，是绸类织品中最紧密的一个品种。塔夫绸分素塔夫、花塔夫、方格塔夫、闪色塔夫和紫云塔夫等多种。花塔夫绸是塔夫绸中的提花织物，地纹用平纹，花纹是八枚缎组织。由于经线分布紧密，因此花纹光亮凸显，质地坚牢、轻薄挺括、色彩鲜艳、光泽柔和，但不宜折叠重压。纹样一般偏中型、大型，花派流畅、大方。塔夫绸花色繁多，色彩鲜艳夺目，有黑、白、粉红、大红、宝蓝、驼色、玫红等，常用于婚纱礼服的面料，如图4-6所示。

Taffeta. Taffeta is a kind of high–grade silk fabric made of plain weave. The warp yarn is made of double–twisted cooked silk, and the silk is made of combined single–twisted cooked silk, with the plain weave as the ground. The large density makes the fabric the most compact one of its kind. There are many kinds of taffeta, such as plain taffeta, pattern taffeta, square taffeta, flash taffeta, and purple cloud taffeta. The pattern taffeta is a jacquard fabric in taffeta, with plain weave as the ground and the eight satin weave as the pattern. Due to the close density of the warp threads, the pattern is highlighted and bright, the texture is firm, light and crisp, the color is bright, and the luster is soft, but it should not be folded and pressed. The patterns are smooth and graceful, and are generally medium–sized and large. Taffeta has a wide variety of colors, including black, white, pink, bright red, sapphire blue, camel, rose red, etc. It is also a common fabric for wedding dresses, as shown in Figure 4–6.

图4-6　塔夫绸面料
Taffeta Fabric

图4-7　真丝绒面料
Real Velvet Fabric

6. 真丝绒面料/Real Velvet Fabric

丝绒是一种既轻又保暖的面料，手感丝滑，柔韧亲肤，与人体有极好的生物相容性。用丝绒面料制作的旗袍更是旗袍中的经典，能够展现东方女子的神韵。丝绒旗袍不但面料看上去比绸缎旗袍更显温柔，而且保暖功能更胜一筹。丝绒面料粗略可见的肌理和质感显得厚重，很适合秋季穿着，如图4-7所示。

Velvet is a lightweight and warm fabric

that feels silky and flexible. It is skin-friendly, with excellent biocompatibility with human body. The cheongsam made of velvet is a classic among cheongsams, and it can best show the charm of oriental women. The velvet cheongsam not only looks much softer than the satin cheongsam, but also has a better thermal function. Its roughly visible texture makes the clothing heavy and warm, very suitable for wearing in autumn, as shown in Figure 4-7.

（二）棉麻类/Cotton and Linen

棉是做旗袍的常用面料，适合制作中西结合式的旗袍，能给人温柔舒适的感觉。老土布、纯棉布、蓝印花布、棉麻混纺布等是日常穿用旗袍的常用面料，简单质朴，深受崇尚自然舒适、喜欢时尚的年轻人的青睐。棉麻面料就是由棉和麻混纺而成的一种面料。做成旗袍既朴素大方，又具有良好的透气性和吸湿性，手感柔软，穿着舒适，凸显少女感，也适合日常穿着。

As a commonly used fabric for cheongsams, cotton is pursued by making cheongsams that integrate both Chinese and Western styles, making the clothing look gentle and comfortable. For example, the old-fashioned cloth, pure cotton cloth, blue printed cloth, cotton and linen blended cloth used for making cheongsam, etc., were commonly used fabrics for daily wearing cheongsam. They are simple and pursued by young people who advocate natural comfort and love fashion. Cotton and linen fabric is a kind of fabric that is blended with cotton and hemp fibers. The cheongsam made in such fabric appears simple and dignified, having good air permeability and moisture absorption. It feels soft, and comfortable to wear, highlighting the temperament of young girls, suitable for daily wear.

在夏季，可选择由纯棉印花细布、印花府绸、色织府绸、什色府绸、各种麻纱、印花横贡缎、提花布等薄型面料制作的短旗袍，轻盈、凉爽、美观、实用。春秋季可选择化纤或混纺面料制作旗袍，如各种闪光绸、涤丝绸，以及各种薄型花呢等面料。这些面料虽然吸湿性和透气性差，但其外观比棉挺括平滑、绚丽悦目，如图4-8和图4-9所示。

In summer, thin fabrics such as pure cotton muslin, printed poplin, yarn-dyed poplin, assorted poplin, various hemp yarns, printed horizontal satin, and jacquard are mostly selected to make short cheongsams, they are light, cool, beautiful and practical. In spring and autumn, chemical fiber or mixed textiles, such as various sparkling silk, polyester silk, and various thin tweed are preferred to make

图4-8　纯棉条纹面料　　图4-9　老土布面料
Cotton Striped Fabric　　Old-fashioned Fabric

cheongsams. These fabrics have poor hygroscopicity and air permeability, yet their appearance is crisper, smoother and more beautiful than cotton fabrics, as shown in Figure 4-8 and Figure 4-9.

（三）毛呢类/Woolen Fabrics

图4-10　毛呢面料
Woolen Fabric

精纺毛织品弹性良好，柔软性和抗皱性好，制作的旗袍结实耐穿、挺括、不易变形。毛呢类面料属于蛋白质纤维制品，光泽柔和自然，手感柔软，比其他天然纤维制品更有弹性，抗折皱性好，熨烫后有较好的褶皱成型和保型性，保暖性好，吸汗及透气性较好，穿着舒适，如图4-10所示。

Refined woolen fabric boasts good elasticity, good softness and strong wrinkle resistance, which allows the cheongsam made in such fabric strong and durable, and not easy to deform. It belongs to protein fiber, has a gentle and natural luster, feels much softer, and is more elastic than other natural fibers. Such fabric is comfortable and easy to be shaped in wrinkles after ironing. It can better keep warm, with good sweat absorption and breathability, as shown in Figure 4-10.

二、旗袍的丝绸面料再造/Reconstruction of Cheongsams' Silk Fabrics

（一）刺绣工艺面料/Embroidery Craft Fabrics

旗袍是东方女性的传统服装之一，能够把东方女性的美刻画得淋漓尽致，并不失传统文化的内涵。旗袍上多以刺绣花卉为主，用刺绣把花卉展现得尽善尽美，体现了中国传统旗袍的韵味，与华贵的传统图案结合在一起，创造出美好的意境。中国传统刺绣元素被运用到旗袍的设计中，使传统文化重放异彩。

Cheongsam is one of the traditional Chinese women's clothing, which has done a good job in presenting the beauty of oriental women incisively and vividly, without losing the connotation of traditional culture. Most of the cheongsams are decorated with embroidered flowers that are displayed perfectly with the embroidery techniques. The flower shapes reflect the charm of traditional Chinese cheongsam, which, coupled with the luxurious traditional patterns, successfully create a beautiful artistic conception. Applying the traditional Chinese embroidery in the design of cheongsam can revitalize the traditional culture.

1. 花卉元素刺绣面料/Floral Embroidery Fabric

在中国的传统服饰刺绣文化中，花卉图案色彩鲜艳，形态万千，代表着吉祥如意，物丰人和，许多花卉图案以不同的形式点缀在旗袍上，使其更加富有设计内涵，并承载着很多美好寓意。比如牡丹纹，牡丹刺绣在服饰上，作为富贵吉祥的象征，表达了人们对幸福的渴求；荷花纹，人们多以写实的手法展现其姿态优美，荷叶疏松有序，寓意爱情、友谊、幸福等美好愿望；梅花纹，表达友情长存，永葆情谊；宝相花纹，代表人们对美好生活的愿望，是富贵吉祥的象征，如图4-11~图4-13所示。

In Chinese traditional clothing embroidery culture, flower patterns are bright in color with shapes, representing good luck, wealth and harmony. Many flower patterns are used in different ways as decorations on the cheongsam, making the clothing more significant in design connotation that bears good meanings. For example, the peony pattern embroidered on clothing as a symbol of wealth and auspiciousness, expresses people's desire for happiness; the lotus pattern presented with realistic techniques to show its graceful posture is a symbol of love, friendship, and happiness; the plum pattern stands for the long-lasting and eternal friendship; the Baoxiang pattern represents people's wishes for a better life and is a symbol of wealth and auspiciousness, as shown from Figure 4-11 to Figure 4-13.

图4-11　牡丹纹　　　　　　图4-12　宝相花纹　　　　　　图4-13　牡丹蝴蝶纹
Peony Pattern　　　　　　　Baoxiang Pattern　　　　　　Peony Butterfly Pattern

2. 动物元素刺绣面料/Animal Embroidery Fabric

刺绣动物图案也是旗袍文化中很重要的一部分。龙凤纹是我国传统装饰纹样中应用最久且广的图案，龙的神奇威武与凤的艳丽美妙构成了人们意念中美好与祥瑞的生动组合，被奉为地位最高的吉祥物。鱼纹在中国人心目中寄托了人们对生活的憧憬，对生活富裕的期盼。鹿纹寓意官吏加官晋爵，官运亨通。这些纹样通过刺绣的形式应用于旗袍的设计中，更能彰显旗袍的华丽与高贵，如图4-14所示。

Embroidered animal patterns are also an important part of cheongsam culture. The dragon and phoenix patterns have remained the longest and most widely used pattern in the Chinese traditional decorative patterns. The magical power of the dragon and the beauty of the phoenix constitute a vivid combination of beauty and auspiciousness in people's minds, thus being regarded as the mascots with the highest position. Fish pattern in Chinese people's mind pinned people's longing for good life and hope to live a rich life. The deer pattern expresses people's desire for promotion and wealth. These patterns are applied to the design of the cheongsam in the form of embroidery to better demonstrate the splendor and nobility of such costume, as shown in Figure 4-14.

3. 风景元素刺绣面料/Landscape Embroidery Fabric

气象与农牧业生产密不可分，因为农牧业生产的好坏关键是气象。古人信奉云神、水神、火神等神灵，祈祷神灵能够适宜地调节气象变化。经过抽象或者具象的形态变化，这些元素完美地出现在刺绣上，如云纹、海水江崖纹、龙袍十二章中的火纹等。这些纹样装饰在旗袍醒目部位，具有象征意义和装饰美感，如图4-15所示。

Meteorology is inseparable from and plays a decisive role in the agricultural and animal husbandry production. The ancients believe in gods such as the god of cloud, god of water, god of fire, etc., and believed gods living amidst clouds could control the change of metrology. Through abstract or figurative morphological changes, these images appear vividly on embroidery. For example, the cloud pattern, the pattern with river and cliff, the fire pattern in the twelve dragon robes, etc., decorated in the eye-catching parts of the cheongsam in abstract images make the clothing have greater symbolic significance and beautiful, as shown in Figure 4-15.

（二）扎染面料/Tie-dye fabrics

扎染古称扎缬、绞缬、染缬等，是中国民间传统而独特的染色工艺。扎缬图案虽然是预先设计好的，但由于扎缬的外力作用，使织物染色不均，拆除扎线后，织物上常会出现一些意想不到的独特花纹。扎染是指织物在染色时，被部分结扎起来，使之不能均衡着色的一种染色方法。扎染布料制作的旗袍，大多色彩斑斓，颇具民族特色。

Tie-dyeing, known as bandhnu, twisting and dyeing in ancient times, is a traditional and unique dyeing process in Chinese folk. The tie pattern is pre-designed, but the dyeing of the fabric is uneven due to the external force of the tie. When the ties are removed, some unexpected and unique patterns appear on the fabric. Tie-dyeing refers to a dyeing method in which the fabric is partially tied when dyeing, so that the clothing cannot be evenly colored. The cheongsams made of

图4-14　凤戏牡丹花纹
Phoenix and Peony Pattern

图4-15　古典云纹与龙纹结合
Combination of Classical Cloud Pattern
and Dragon Pattern

tie-dye fabrics are colorful with distinctive ethnic characteristics.

在扎染工艺中，扎缬和染色作为两个不同的工艺概念，在工艺流程中，顺序是先扎后染，而扎缬的质量是决定最后染色效果的关键因素。不同的扎缬方法设计出的纹样图案也不尽相同。扎染的本质是"防染"，利用这一本质，就能够不断发现新的扎缬材料和扎缬方法，使扎染作品充满独特性，这正是扎染技艺的独特魅力。天然染料蓝靛是染色工艺的主要染料，也可以用化学染料进行染色。根据预期的效果，在染色前做好安排，无论是染色时间的人为控制，抑或多种颜色的染色顺序先后，最后都会产生不同的、出乎意料的扎染效果。不管是传统的纹样，还是现代的几何、花卉纹样，这些独具一格、美轮美奂的纹样和多变的晕染效果，都会给人带来不同的感受和丰富的联想。

In the tie-dyeing process, tying and dyeing are two different concepts, that is, dyeing comes after tying. And how the fabric is tied determines the final dyeing effect. Different tying methods result in different patterns. The essence of tie-dyeing is "anti-dyeing", according to which, new tie-dye materials and tie-dye methods can be continuously discovered to create unique tie dyeing works. This is the exclusive charm of tie-dye techniques. The natural dye indigo acts as the basic material for the dying technology, of course, chemical dyes can also serve for this purpose. Given the expected effect, prior arrangements before dyeing are made, be it the artificial control of the dyeing time, or the dyeing sequence of multiple colors, to make different and unexpected tie-dyeing effects. Whether it is traditional patterns or modern geometric and floral patterns, these unique and beautiful patterns, together with the changing effects, will bring people different feelings and enrich their imagination.

将旗袍技艺与扎染工艺进行创新设计，可以赋予植物染新的面貌，将扎染工艺与旗袍结合起来，明暗层次的对比，简单、原始的蓝白两色，创造出一个淳朴自然、千变万化、绚丽多姿的蓝白艺术世界，反映一种深厚的文化和艺术积淀，如图4-16和图4-17所示。

Carrying out innovative design on the cheongsam technique with the tie-dye technique has injected the plant dying technology with new energy, and created a simple, natural, ever-changing and gorgeous world with the original blue and white color in strong contrast of light and dark, thereby reflecting a profound culture and art, as shown in Figure 4-16 and Figure 4-17.

（三）数码印花工艺面料/Digital Printing Fabrics

数码印花是用数码技术进行的印花。数码印花技术是随着计算机技术不断发展而逐渐形成的一种集机械、电子信息技术于一体的高新技术产品。

Digital printing refers to the printing using digital technology. It is a high-tech product that integrates machinery, and computer electronic information technology with the continuous development of computer technology.

数码印花印制的花纹具有精细、明晰、层次丰厚、自然的特性，能够印制类似于照片和绘画风格的产品。数码印染是一种全新的印花方式。利用数码喷墨印花机把染料直接喷射在面料上，快速形成高精度的彩色图案。数码印花一般用于批量制作，具有速度快、精准度

图4-16 扎染面料
Tie-dye Fabrics

图4-17 扎染旗袍
Tie-dye Cheongsam

图4-18 喷绘旗袍
Inkjet Cheongsam

高的特点。这种艺术与技术的结合给人一种性感、时尚的感觉。利用数码印花方式生产的面料制作的旗袍，图案更加丰富多彩，打印速度快，适合小批量生产，更能彰显旗袍的美感和个性，因而广泛应用于旗袍面料的生产，如图4-18所示。

The patterns printed by digital printing are fine, clear, rich in layers, and natural, and can print products similar to photos and paintings. As a brand-new way of printing, the technology uses a digital inkjet printing machine to spray directly the dye on the fabric to quickly form a high-precision color pattern. Generally, such printing and dyeing usually serve batch production due to its characteristics of high speed and high precision. This combination of art and technology makes the fabric sexy and fashionable. The cheongsam made of fabrics produced by digital printing and dyeing is of more colorful patterns that are printed at a fast speed, suitable for small batch production, showing its beauty and personality in a better way. Therefore, it is widely used in the production of cheongsam fabrics, as shown in Figure 4-18.

（四）手绘工艺面料/Hand-painted Craft Fabrics

手绘服装，通俗地讲就是在衣服上画画，最早源于美国的街头艺术。手绘采用的颜料是从植物中提取的，具有防水功能。无论是机洗还是手洗，都不会洗掉色，而且颜料对人体也没有任何伤害。旗袍的手绘图案，可以是花鸟鱼虫、山水风景，也可以是独特的定制图案。这样的手绘旗袍不仅独一无二，可以更好地凸显自己的气质，而且具有收藏价值。

Hand-painted clothing, generally speak, is to draw on clothes, which originated from street art in the United States. The pigments used are extracted from plants and are waterproof with no harm to human body. Whether it is washed by machine or by hand, the pattern will not be washed off. The hand-painted patterns of the cheongsam can be flowers, birds, fish and insects, landscapes, or unique custom patterns. Such hand-painted cheongsam not only can better highlight its unique temperament, but also has collection value.

手绘特点是随需而绘，不同于刺绣、印染、丝网印刷等，它的开发潜力大，采用专业的植物颜料，具有颜色浓淡、层次变化丰富的特点，线条晕开的形态是机器达不到的视觉效果。手绘面料手感好，能保持衣服的柔软性。手绘与绣花、缝珠片、烫钻、珍珠、亮片等互相搭配，更能凸显图案的立体感，视觉效果好，符合崇尚个性的潮流，如图4-19、图4-20所示。

Hand-painting is different from embroidery, printing and dyeing, and screen printing. It has great potential for development, using professional plant pigments with good visual effects and diverse changes that cannot be realized by machines. Hand-painted clothing feels soft and can maintain the softness of the clothes. Matching such fabric with embroidered flowers, sewn sequins, hot diamonds, pearls, sequins, etc. can better highlight the three-dimensional sense of the pattern to create splendid visual effect, which is in line with the trend of advocating individuality, as shown in Figure 4-19 and Figure 4-20.

图4-19　丝绸旗袍面料上的手绘图案
Hand-painted Pattern on Silk Cheongsam

图4-20　手绘荷花旗袍
Hand-painted Lotus Cheongsam

第二节　丝绸旗袍图案风格/Pattern Style of Silk Cheongsam

经过数百年的演变，旗袍上的图案也越来越讲究，越来越强调"图案装饰"的时尚作用。旗袍图案或面料本身的图案，或单独镶嵌的图案，或利用其他面料剪裁的图案，或特殊制作的图案，都是为了点缀旗袍而设计的。没有图案的旗袍就略显呆板肃穆，而有图案装饰的旗袍就显现多种风格，彰显高雅端庄、雍容华贵、美丽大方、浪漫飘逸等不同气质。

After hundreds of years of evolution, the patterns on cheongsam are becoming more and more particular, placing increasing emphasis on the fashionable role of "pattern decoration". The cheongsam pattern or the pattern of the fabric itself, or the pattern inlaid alone, or the pattern cut from other fabrics, or the specially made pattern, are all designed to embellish the cheongsam. A cheongsam without a pattern is a little dull and solemn, while a cheongsam decorated with a pattern shows a variety of styles, becoming elegant, dignified, graceful, romantic and gorgeous.

面料本身的图案也是配合旗袍款式色彩、面料而进行装饰的。传统织锦缎面料一般都是大圆的图案和团花图案，为了利用它本身的图案需进行巧妙的安排。素色面料在装饰图案时采用刺绣、印花、绘画等装饰艺术手段，起到画龙点睛的作用。

The original pattern of the fabric is also decorated with the style, color, and fabric of the cheongsam. For traditional satins and brocades that have large circle patterns and round patterns, clever cutting is needed to make good use of these patterns. For plain fabrics, embroidery, printing, painting, and other decorative means are used to decorate the pattern and make the clothing more beautiful.

清朝旗装过分注重装饰细节，衣身色彩面积大且较为艳丽，对比度和视觉冲击感较强，图案繁杂。袖襟饰边纹样的样式变化大，与衣身的整体图案较为融合。在清朝旗装的装饰图案中，又以团花纹样最为普遍。清末年间的旗袍图案以团花、万字纹、水纹为主，体现吉祥如意。

The Banner gowns in the Qing Dynasty paid too much attention to decorative details, with large and bright colors, and complicated patterns, creating a strong contrast visual impact. The patterns of the sleeves, Jin, and hems vary greatly, getting integrated with the overall pattern of the clothing. Among the decorative patterns of the Banner gowns in the Qing Dynasty, the group pattern was the most common. The cheongsam patterns in the late Qing Dynasty were mainly composed of round patterns, swastikas, and water patterns, to reflect good luck and wishes.

民国早期，旗袍图案的实现手法与清代服饰图案的实现手法相似，通常以提花织造和手工刺绣为主，旗袍装饰图案以自然图案和动物图案为主，常见的有龙凤、梅兰竹菊、蝶鹤百鸟等纹样，通过相互穿插叠放，或重复循环的方法组合成不同的单一或复杂图案。

In the early period of the Republic of China, the creation method of the cheongsam pattern was virtually similar to that of the Qing Dynasty's clothing, centering on jacquard weaving and hand embroidery. Natural and animal patterns dominated the decorative patterns of the cheongsam.

The common ones were dragon and phoenix, plum, orchid, bamboo, chrysanthemums, butterflies, cranes and birds, which were combined into different single or complex patterns by interspersed and stacked with each other, or in repeated forms.

20世纪20年代，旗袍领、袖、襟都装饰连续花边图案，而且层次密集，体现了花边装饰美。30年代以后，旗袍图案变化非常丰富，受西方文化的影响，西化的图案比较多，条格纹、碎花纹的构成都流行过，团花图案重新兴起，团花之间常常用藤蔓、枝叶等纹样来联系填充，讲究与主体图案在色彩与形态上的映衬。团花图案常给人丰富绚丽的感觉，寓意大富大贵，符合传统的审美心理。

In the 1920s, the collar, sleeves, and front of the cheongsam were decorated with continuous lace patterns with multiple layers, reflecting the beauty of lace decoration. After the 1930s, the pattern of cheongsam changed significantly. Influenced by Western culture, there were many westernized patterns, such as striped plaid and broken patterns. The round pattern was revitalized, which was connected and filled with vines and leaves, to set off the main pattern both in color and form, creating a feeling of splendid and gorgeous, implying great wealth and honor, and conforming to the traditional aesthetic psychology.

现代旗袍图案设计，首先要考虑旗袍的设计风格、款式造型的变化、点缀部位。位置确定之后，需要考虑用什么形式装饰图案，是拼接异色还是刺绣图案或镶嵌图案。

In the design of modern cheongsam patterns, the design style of the clothing should be firstly considered, followed by the changes in style and shape, and the embellishment parts. After the location is determined, it should be considered what form of decorative pattern should be used, stitching colors, embroidering patterns, or mosaic patterns.

现代旗袍受到了西式理念的影响，接收了许多西方时尚潮流，旗袍的款式和花纹样式也逐渐变多，由传统的花纹变成丰富多彩的印花样式。随着生产技术的发展，织花、印染、刺绣等工艺技术的应用，现代旗袍出现了更多精彩的装饰纹样。

Modern cheongsam has been influenced by Western concepts and Western fashion trend ideas, and its styles and patterns of cheongsam have gradually increased, changing from traditional patterns to colorful printing styles. With the development of production technology and the application of weaving, printing and dyeing, embroidery and other craftsmanship, modern cheongsams have more wonderful decorative patterns.

中国服饰的传统纹样主要有九大类：龙蟒、凤凰、珍禽、瑞兽、花卉、虫鱼、人物、几何与寓意。

The traditional patterns of Chinese clothing mainly fall into nine categories: dragon, phoenix, rare birds, auspicious animals, flowers, insects and fish, figures, geometry and symbolic patterns.

一、"卍" 字纹 / "卍" Pattern

"卍"字纹是一种宗教标志，这种标志旧时译为"吉祥海云相"。在武则天当政时，被正式用作汉字。象征光明，还有轮回不止的意思，与太极图有异曲同工之妙。"卍"字符寓意鱼水共欢、阴阳和合、富裕寿喜、吉祥如意。

The "卍" pattern, a religious symbol, meant "auspicious sea and clouds" in the old days. When Wu Zetian (the only female emperor in China in the Tang Dynasty) was in power, it was officially used as a Chinese character. It symbolizes light and eternity, similar to Tai Chi. The character of "卍" stands for the common joy of making love, the harmony of yin and yang, prosperity and longevity, and good luck.

我国出现"卍"字纹可以追溯距今4000多年的新石器时代晚期的马家窑文化。新石器时代，陶器、古巴蜀国的铜带钩、唐代铜器、清代织锦、镂空门窗上比比皆是。这些器物上使用的"卍"字纹大多是取吉祥寓意，审美成分越来越浓，渐渐演变成民族传统的审美对象，寓有风调雨顺、万寿无疆之意，如图4-21、图4-22所示。

The appearance of the "卍" pattern in China can date back to the Majiayao culture in the late Neolithic Age more than 4, 000 years ago. It can be seen in the Neolithic pottery, copper belt hooks from the Shu Kingdom in the Ancient Shu Dynasty, bronze wares in the Tang Dynasty, brocades in the Qing Dynasty, and hollow doors and windows. Most of this patterns used on these utensils have auspicious meanings with increasing aesthetic significance, and gradually evolved into national traditional aesthetic objects that imply good weather and longevity, as shown in Figure 4-21 and Figure 4-22.

图4-21 红地"卍"字纹缎夹旗袍（中国丝绸博物馆）
Satin Cheongsam with "卍" Pattern on Red Cloth
（China National Silk Museum）

图4-22 宋锦"卍"字纹中装（2014年APEC领导人会议服装）
"卍" Pattern Chinese Clothing Made in Song Brocade（2014 APEC Leaders' Meeting Clothing）

二、水波纹与云纹/Water-ripple and Cloud Patterns

（一）水波纹/Water-ripple Pattern

水波纹又称"海涛纹""海水纹"。水滋养万物，造福万物，因此水波纹被世人赋予了厚德载物、海纳百川的寓意。

Water-ripple pattern is also known as "sea wave pattern" and "sea water pattern". Water nourishes and benefits all things. Therefore, the water-ripple patterns are endowed with the meaning of great virtue and greatness, as inclusive as the sea that can accommodate all rivers.

波涛汹涌的水波纹有气势磅礴、威严不可触犯之态，多用于龙袍、蟒袍、官府的袖口、下摆，常与龙纹、蟒纹、凤纹等相配使用，是身份的象征。由水波纹与山石组合而成的"海水江崖纹"象征江山，有一统江山，福山寿海，江山永固之意。"立水"指袍服最下摆条状斜纹所组成的浪水；"平水"指在江崖下面如鳞状的海波。海水意指"海潮"，"潮"与"朝"同音，故成为官服的专用纹饰；"江崖"即山头重叠，似姜之芽，象征山川昌茂、国土永固。

The turbulent water-ripple patterns have a majestic and awe-inspiring appearance that can not be offended. It is mostly used with the patterns of dragon, python and phoenix on the cuffs and hems of dragon robes, python robes, and official wears as the symbol of the wearers' identities. The "sea water river and cliff pattern", which is composed of water ripples and rocks, symbolizes the power, implying the firm ruling power of a country and long lifetime. "Lishui" refers to the wave water formed by the striped twill at the bottom of the robe; "Pingshui" refers to the scaly sea waves under the river cliffs. Sea water means "ocean tide", and the tide sounds similar to the official government in Chinese, therefore becoming a special decoration for official uniforms; "Jiangya" is the overlapping of hills, stands for the prosperity and the eternal stability of the country.

除了与龙纹样组合，海水江崖纹也可与其他寓意吉祥的纹饰搭配，如蝙蝠、葫芦、祥云、八宝纹等。水波纹图案寓意吉祥，色彩丰富，造型唯美，被广泛应用于旗袍和现代服饰中，如图4-23、图4-24所示。

In addition to being combined with the dragon pattern, the "sea water river and cliff pattern" can also be matched with other auspicious patterns, such as bats, gourds, auspicious clouds, and Babao pattern. The water ripple pattern that implies auspiciousness is rich in color and beautiful in shape, enjoying wide application in cheongsam and modern clothing, as shown in Figure 4-23 and Figure 4-24.

图4-23　水波纹在龙袍中应用
Application of Water-ripple Pattern in Dragon Robes

图4-24　水波纹在官服中应用
Application of Water-ripple Pattern in Official Uniforms

（二）云纹/Cloud Pattern

云纹是我国丰富多彩的装饰纹样中的一种，是非常典型的古代吉祥图案，象征吉祥、如意，造型优美，变化多样，被广泛应用在中国古代的服饰、建筑、器具及各种工艺品上。

As one of the extensive decorative patterns in China, the cloud pattern is a very typical ancient auspicious pattern that symbolizes good luck and wishes with beautiful shape and diverse changes. It is widely used in ancient Chinese clothing, architecture, utensils and handicrafts.

隋唐是云纹盛行的时期。宋代云纹总体上依然是朵云样式。由于古代长期的采集和耕作实践，人们对云和雨决定收成的影响产生期盼和敬畏，因此云在人们心中的形象得到升华和抽象。唐宋以来，为顺应时代的审美要求，装饰元素日趋丰富，在中国云纹体系中，如意云纹是最具抽象品格、又被普遍认同、应用广泛的一种类型。

The Sui and Tang Dynasties witnessed the popularity of the cloud pattern which was kept in primary use in the Song Dynasty. Due to the long-term collection and farming practices in ancient times, people at that time showed great respect to the fact that clouds and rains played a decisive role in the agriculture industry, and placed their good wishes and respect on them. Since the Tang and Song Dynasties, the decorative elements have become increasingly extensive in line with the aesthetic needs of the times. In the Chinese cloud pattern system, the Ruyi cloud pattern is the most abstract one with the widest recognition and application.

经过几千年的发展演变，云纹图案的种类更加丰富，每个历史时期的云纹样式都融入了各自时代的不同风貌。云纹形态多样，有十分抽象规则的几何图形，也有生动形象的自然图形，造型优美，形态多样，寓意吉祥，在旗袍等服饰中应用较广泛，如图4-25～图4-27所示。

After thousands of years of development and evolution, the kinds of cloud pattern have been enriched. And the pattern of each historical period was based on the different styles and features of its own era. There are various forms of cloud patterns, including very abstract and regular geometric figures, and vivid natural figures. They are beautiful and diverse in shape as the symbol of auspiciousness and are widely used in cheongsam and other clothing, as shown from Figure 4-25 to Figure 4-27.

三、动物纹样/Animal Patterns

在旗袍设计中动物纹样应用最多的是龙凤纹样，象征祥兆。因为过去有"龙凤呈祥"的说法，所以人们认为龙凤寓意一生富贵。此外，还有一些蝴蝶仙鹤等动物图案也得到了应用。如旗袍图案中常使用龙与祥云，龙与凤的结合，这两种图案在早期旗袍设计中是比较受欢迎的，且具有中华古典韵味。

The dragon and phoenix patterns are the most used animal patterns in the cheongsam design as the symbol of auspiciousness. Because of the saying that "the dragon and the phoenix are auspicious" in the past, people think that the two animals stand for wealth and honor in life. In

图4-25 水波纹在现代
服饰中的应用
The Application of Water-
ripple Pattern in Modern
Clothing

图4-26 云纹在现代服
饰设计中的应用
The Application of
Cloud Pattern in Modern
Clothing Design

图4-27 云纹与水波纹的
在旗袍中的应用
The Application of Cloud
Pattern and Water-ripple
Pattern in Cheongsam

addition, some animal patterns such as butterflies and cranes have also been applied. For example, the patterns of cheongsam are commonly used the combination of dragon and auspicious cloud, dragon and wind. These two patterns became popular in early cheongsam with profound Chinese classical charm.

(一) 龙纹/Dragon Pattern

龙的纹样贯穿着中华历史，而龙的形态具有极强的亲和力，它本身不仅是多种动物的结合体，而且造型的适应性也很强，可与多种几何纹样较好地相融。比如原始时期的龙与S纹、涡纹结合，秦汉时期的龙与云纹、气纹大量结合，战国时期龙凤虎纹刺绣，佛教的传入又涌现了"如意龙纹""方胜龙纹"等。

The pattern of the dragon runs through Chinese history. The shape of the dragon is full of affinity. It's not only a combination of various animals, but also changing in its shape, which can be well integrated with a variety of geometric patterns. For example, the dragon in the primitive period was combined with the S pattern and the vortex pattern, and with a large number of cloud patterns and air patterns in the Qin and Han Dynasties. There was the dragon, phoenix and tiger pattern embroidery in the Warring States Period, and the introduction of Buddhism gave birth to the "Ruyi dragon pattern" and "Fang Sheng (a pattern of two diagonally overlapping squares) dragon pattern".

龙纹表示风调雨顺、吉祥安泰和祝颂平安与长寿之意。凤纹象征着吉祥、勇敢、神力、希望以及美好的爱情。

The dragon pattern represents good weather, luck, and wishes for peace and longevity, and the phoenix pattern symbolizes good luck, bravery, divine power, hope and beautiful love.

在岁月的长河中，我们的祖先创造了许多向往美好生活、寓意吉祥的独具民族特色的图案。中国服饰的传统纹样图案巧妙地运用花卉、虫鱼、凤凰、龙蟒等，表达人们对于美好事物的追求。

Over the years, our ancestors have created many patterns with unique national characteristics that imply our struggle and pursuit for a better life and good wishes. Such pursuit for beautiful things has been well expressed through the traditional patterns of Chinese clothing with patterns of flowers, insects, fish, phoenix, dragon, and python.

婚礼或者礼服类旗袍，龙凤图案出现很多。但如果是非婚礼旗袍，龙凤只是作为次要的装饰，并用花草加以点缀，弱化了龙凤的象征意义。和花草相配合出现的喜鹊等鸟类也是旗袍中常出现的纹样，如图4-28~图4-30所示。

There are many applications of dragon and phoenix patterns for weddings or dress cheongsam. However, for a non-wedding cheongsam, the dragon and phoenix come as the less important decorations, and will be embellished with flowers and plants to weaken the symbolic meaning of the dragon and the phoenix. Other common patterns in cheongsams include birds such as magpies in combination with the fragrance of flowers and plants, as shown in Figure 4-28 to Figure 4-30.

图4-28　龙袍
Dragon Robe

图4-29　龙纹旗袍
Dragon Pattern Cheongsam

图4-30　Ralph Lauren 2011秋冬系列中旗袍裙背后的四爪蟒纹
The Four-clawed Python Pattern on the Back of the Cheongsam in the Ralph Lauren 2011 Autumn and Winter Series

（二）凤纹/Phoenix Pattern

凤纹是中国具有悠久历史的传统装饰纹样之一，在传统服饰中广泛使用，具有丰富而多样的文化内涵。凤纹因其美好寓意在现代旗袍设计中广泛应用。

The phoenix pattern is one of the most traditional decorative pattern with a long history in China. It has been widely used in traditional clothing with rich and diverse cultural connotations, as well as in the design of modern cheongsam because of its beautiful meaning.

凤是由自然界中各种不同动物融合而成的神物。凤与龙一样，在经历了几千年的历史文化的洗礼延续至今，成了中华民族最具生命力的文化标志和精神象征。虽经过不同时代的演变，崇拜凤的观念与日俱增，逐渐成为人们心中的意识形态。

The phoenix is actually a fetish created by the fusion of various animals in nature. Like the dragon, the phoenix has become the most vital cultural and spiritual symbol of the Chinese nation after thousands of years of history and culture development. Chinese people's worship of the phoenix has become greater despite its evolution throughout different times, making it an ideology in our hearts.

魏晋南北朝时期，凤纹的形式特征在汉代的基础上进一步深化发展，由于当时佛教盛行，凤鸟形象被注入了新的精神理念。花卉缠枝纹样被广泛应用，凤多采用在清新的花卉图案之中展翅飞翔的形象。到了明代，凤纹已经成为一种特定的造型，纹样构成都各具其内在形式，凤纹的共性形态也进一步规范化。清代纹饰以凤纹服饰较明代增多，而且画法风格各不相同，按纹饰的组成分为双凤、团凤、夔凤、凤凰牡丹纹、龙凤纹等。现代凤纹因其象征吉祥，被广泛应用于结婚礼服和旗袍设计中，如图4-31、图4-32所示。

During the Wei, Jin, Southern and Northern Dynasties, the formal characteristics of the phoenix pattern were further developed on the basis of the Han Dynasty. Due to the prevalence of Buddhism at that time, the image of the phoenix was injected with new spiritual concepts. Accompanied by the extensive application of the flower and branch patterns, the most common image of this pattern is a phoenix that is spreading its wings and flying among the fresh flowers. By the Ming Dynasty, the phoenix pattern has become a specific shape, and the composition has its own internal form. The common form of phoenix pattern got further standardized. The number of costumes decorated with phoenix patterns in the Qing Dynasty increased more than that in the Ming Dynasty, and the styles of painting were different. Modern phoenix patterns are widely used in wedding dresses and cheongsam designs

图4-31　郭培玫瑰坊纯手工刺绣旗袍
Guo Pei Rose Studio Hand-Embroidered Cheongsam

because of their auspicious symbols, as shown in Figure 4–31 and Figure 4–32.

图4-32　凤纹在旗袍上的运用
The Application of Phoenix Pattern in Cheongsam

(三) 孔雀等其他动物纹样/ Peacock and Other Animal Patterns

孔雀纹是深受民间喜爱的纹样，孔雀在传说中是拥有九德的吉祥鸟，由于孔雀尾羽纹饰美丽，开屏时"纹饰明显"，和"文明"谐音，故孔雀开屏则寓意"天下文明"，表示人们对盛世的向往。

Peacock pattern is popular among the common people, and peacock has nine virtues in the legend. Because of the beautiful decoration of peacock tails and feathers, when a peacock spreads its tails in the pattern, it symbolizes civilization in Chinese, referring to the yearning for a prosperous society.

虎纹在所有的生肖中，几乎是最具时尚神韵的动物，霸气、骄傲、神秘、性感、时尚，在现代旗袍设计中体现了刚柔相济的美感，如图4-33所示。

Tiger pattern is almost the most fashionable animal among all the zodiac signs. It is domineering, proud, mysterious, sexy, and fashionable, reflecting the beauty of hardness and softness in the design of modern cheongsam, as shown in Figure 4–33.

四、花草纹/Flower and Plant Pattern

花草图案是旗袍纹样当中运用最多的花纹之一，既可以简单简约，又可以富丽华贵。在明清时期，植物纹样在服饰上的应用非常多。民国时期，花卉图案更是成为最受欢迎的装饰图案之一。中国传统的装饰手法多为象征、隐喻、借代、谐音等，借助自然事物装饰服装又有其他意义。人们采用松、竹、梅、菊等植物花卉纹样作为装饰图案，一方面是为了使旗袍整体富有活力，另一方面则是通过图案来隐喻自身的高尚品格。

图4-33　孔雀、虎纹等在现代旗袍中的应用
Application of Peacock, Tiger and Other Animal Patterns in Cheongsam

The flower and plant patterns are one of the most used ones in the cheongsam pattern, which can be simple and luxurious. Plant patterns enjoyed extensive application in clothing during the Ming and Qing Dynasties, while in the Republic of China, floral patterns became one of the most popular decorative patterns. With the methods of symbols, metaphors, metonymy, and homonyms, the traditional Chinese decorative techniques intend to decorate clothing with natural things that have symbolic meanings. For example, pine, bamboo, plum, chrysanthemum and other plant and flower patterns were used as decorative patterns to make the cheongsam energetic and imply the noble characters in the meantime.

随着思想文化的改变，一些没有特别含义的植物也出现在服装上。在此类植物图案中，又以花卉图案居多，如牡丹花卉图案皆有枝叶作为联系，整体自然不浮夸，花瓣以亮色为主，抢人眼目。牡丹寓意大富大贵，深受人们喜爱。

With the change in thinking and culture, some plants with no special implied meanings also appear on clothing, most of which fall into flower patterns. For instance, the peony flower patterns are all connected with branches and leaves to present natural and harmonious beauty, and the petals are so bright to be eye-catching. As the metaphor for great wealth and honor, the peony has been much favored by people.

（一）梅、兰、竹、菊花草纹样/Pattern of Plum, Orchid, Bamboo, and Chrysanthemum

每个时代都有每个时代特有的装饰纹样，在民国时期，旗袍作为主流服饰，它们的装饰纹样，毋庸置疑就是旗袍的基本纹样，是来自中国传统文化中常用到的象征符号，比如花草类的梅兰竹菊，动物类的龙凤鸟兽。早期清代旗袍，还没有受到汉文化的过多影响，仍旧保留着民族的旗袍纹样的特色，简单朴素，几乎没有多余的纹样装饰。

Each era has its unique decorative patterns, and in the period of the Republic of China, the decorative patterns were undoubtedly the basic patterns of cheongsam which were the mainstream clothing style then, coming from the symbols commonly used in traditional Chinese culture, such as the plum, orchid, bamboo and chrysanthemum of flower and plant patterns, and the dragon, phoenix, bird, and beast of the animal patterns. The Banner gowns of the early Qing Dynasty cheongsam had not been influenced too much by the Han culture, and remained the characteristics of its ethic features that were simple with no undue decoration.

在汉文化的影响之下，旗袍的纹样渐渐增多，开始使用大量秀丽的花草图案。梅兰竹菊是最能代表中国传统文化的花草图案，再有就是莲花等具有美好象征意义的花纹图案，如图4-34、图4-35所示。

The patterns of cheongsam got enriched under the influence of Han culture, and started to use many beautiful flower and plant patterns. The patterns of plum, orchid, bamboo, and chrysanthemum could best represent Chinese traditional culture, followed by lotus with good implied meanings, as shown in Figure 4-34 and Figure 4-35.

图4-34 梅、兰、竹、菊纹样在旗袍上的运用
Plum, Orchid, Bamboo and Chrysanthemum Patterns on Cheongsams

1. 梅花/Plum

梅花纹样有五瓣，象征福禄寿喜财；梅花和喜鹊一起出现时，又代表着喜上眉梢的吉祥寓意。梅花有不屈不挠、坚强不屈的精神和顽强意志。它高洁、坚强、谦虚、不与世俗同流合污的品格，历来为我国人民所喜爱，也给人以发奋图强的激励。梅花在传统文化中是高洁、傲骨的象征。

The plum blossom pattern has five petals, which symbolize good luck, fortune, longevity, happiness, and wealth; when plum blossoms and magpies appear together, they imply good things about to happen. Plum has been endowed with an indomitable, unyielding spirit and tenacious will. Its noble, sheer, humble, and arrogant characters, allowing it to be loved by Chinese people for all the time, inspiring them to work hard and become stronger. Plum blossoms are a symbol of nobleness and arrogance in traditional culture.

2. 兰花/Orchid

兰花清新淡雅，深受文人雅士的喜爱。在旗袍上以兰花作为装饰，也能够凸显女性的温婉气质。

图4-35　菊花、兰花纹样创新应用
Innovative Application of Chrysanthemum and Orchid Patterns

Orchids are fresh and elegant, and very popular among refined scholars. Orchid decoration on the cheongsam can also highlight the gentle temperament of women.

3. 竹/Bamboo

高风亮节，坚韧不拔，是竹给人的第一印象。旗袍纹样中的竹的形象，大多是翠绿的竹叶点缀在旗袍上。

Bamboo stands for noble character and sterling integrity, mostly appearing in emerald green bamboo leaves on the cheongsam.

4. 菊花/Chrysanthemum

菊花的花型十分饱满，虽然菊花给人的印象属于悠闲恬淡，但是在旗袍上的菊花图案却有着富贵长寿的寓意。

The chrysanthemum flower is in full bloom, representing the leisure and tranquil life, while having the meaning of wealth and longevity on cheongsam.

（二）牡丹纹样/Pattern of Peony Flower

牡丹雍容华贵，国色天香，是富贵和美丽的象征，是我国十大传统名花，被尊为"花王""国花"。以牡丹花为主调的吉祥图案具有浓郁的中华民族特色，在现代旗袍设计中，牡丹花纹图案应用最为广泛，如图4-36所示。

The peony is graceful and luxurious, and the national color is fragrant. It is a symbol of wealth and beauty, and it is one of the top ten traditional famous flowers in my country. The auspicious pattern with peony as the main tone has strong Chinese national characteristics. In the design of modern cheongsam, the peony pattern is the most widely used, as shown in Figure 4-36.

（1）花开富贵纹样。五代时期，牡丹被赋予"富贵"品格，大画家徐熙作有《玉堂富贵图》。宋代，周敦颐写道："牡丹，花之富贵者也。"盛开的牡丹组图，寓意"富贵吉祥，繁

图4-36 牡丹纹样在旗袍中的运用
Application of Peony Pattern in Cheongsam

荣昌盛，幸福美满"。

Flowers bloom with rich and noble patterns. During the Five Dynasties, peony was endowed with the character of "wealth and honor", and the great painter Xu Xi made *The rich and honorable picture of Yutang*. In the Song Dynasty, Zhou Dunyi said："Peony, the rich and noble of the flower." The picture of the peony in full bloom means "wealth and auspiciousness, prosperity and happiness."

（2）玉堂富贵纹样。玉兰花和海棠简称"玉棠"，与"玉堂"同音双关。

Yutang rich and noble patterns. Magnolia and Begonia are referred to as "Yutang", which is homophonic with "Jade Hall".

（3）凤穿牡丹纹样。晚唐诗人皮日休诗曰："落尽残红始吐芳，佳名唤作百花王。"宋代诗人杨万里诗曰："东皇封作万花王，更赐珍华出尚方。"东皇是民间传说中的司春之神，管理百花，可见牡丹是人民群众公认的"花王"。凤凰是"百鸟之王"，牡丹是"百花之王"，牡丹与凤凰组图，寓意"吉祥喜庆，婚姻美满"。

The phoenix wears the peony pattern. In the late Tang Dynasty, poet Pi Rixiu's poem said："After all the redness has fallen, it begins to spit out fragrance, and the good name is called the King of Hundred Flowers." In the Song Dynasty, Yang Wanli's poem said："The Emperor of the East was named the King of Wanhua, and he gave precious flowers to the Shangfang." Dongguan is the god of spring in folklore and manages hundreds of flowers. It can be seen that peony is recognized by the people as the "King of Flowers". The phoenix is the "king of birds" and the peony is the "king of hundreds of flowers".

（4）缠枝牡丹纹样。武则天贬牡丹的传说在我国影响深远，其实这是艺术虚构，人民群众借牡丹故事表达了中华民族不畏权贵、生生不息的伟大民族精神。缠枝牡丹纹图是传统吉祥纹样，又名"万寿藤"，因结构连绵不断，寓意"生生不息，富贵绵长"。

The pattern of tangled peony. The legend of Wu Zetian's devaluation of peony has far-reaching influence in our country. In fact, it is an artistic fiction. The people use the story of peony to express the great national spirit of the Chinese nation, which is not afraid of the powerful and endless. The pattern of tangled branches of peony is a traditional auspicious pattern, also known as the "Longevity Vine". Because of its continuous structure, it means "endless life, prosperity and longevity".

（5）富贵平安纹样。中国有竹报平安之说，牡丹与翠竹组图，寓意"富贵平安"。花瓶插牡丹，花瓶画牡丹，也寓意"富贵平安"。

Patterns of wealth and peace. There is a saying in China that bamboo is safe, and the peony and green bamboo are pictures, which means "wealth and peace". Putting peonies in the vase and painting peonies in the vase also means "wealth and peace".

（6）长命富贵纹样。牡丹与寿石、桃花组图，或牡丹与寿石组图，寓意"长命富贵，福禄长寿"。

Patterns of longevity and wealth. The picture of peony and longevity stone, peach blossom, or the picture of peony and longevity stone, means "long life, wealth, prosperity and longevity".

（7）牡丹玉兰纹样。牡丹与玉兰组图，寓意"玉堂富贵"。

Peony magnolia pattern. A picture of peony and magnolia, which means "Jade Hall is rich and noble".

（8）牡丹锦鸡纹样。锦鸡为祥瑞之鸟，"鸡"与"吉"同音双关，牡丹与锦鸡组图，寓意"富贵吉祥"。

The pattern of the golden pheasant on the peony. The golden pheasant is the bird of auspiciousness, with the homophonic pun of "chicken" and "auspicious".

（9）花好月圆纹样。在古代，结婚证书也称龙凤证书、富贵证书、良缘证书等，常采用牡丹图案。一轮明月配牡丹、月季、菊花，寓意"花好月圆，家庭美满"。

Flowers are full of moon circle patterns. In ancient times, marriage certificates were also called dragon and phoenix certificates, wealth certificates, and marriage certificates, etc., often using peony patterns. A bright moon is paired with a peony, rose, and chrysanthemum, which means "the flower is full of flowers and the moon is full, and the family is happy".

（10）金鹊报喜纹样。喜鹊或登于牡丹花枝，或飞于牡丹花间，寓意"金鹊报喜，富贵吉祥，婚姻美满"。有时也采用传统团花设计，配桃花、芍药、梅花等名花。

Golden magpie announcement pattern. Magpies either climb on the peony branches or fly among the peony flowers, which means "golden magpies announce good news, wealth and auspiciousness, and a happy marriage". Sometimes the traditional round pattern is also used, with famous flowers such as peach, peony, and plum.

（11）双凤朝牡丹纹样。汉代刘安《淮南子》一书中开始称凤凰为祥瑞之鸟，雄曰凤，雌曰凰。龙凤都是人们心中的祥兽瑞鸟，哪里出现龙，哪里便有凤来仪。牡丹是万花之王，双凤朝牡丹寓意"龙凤呈祥，天下太平，富贵吉祥"。

Shuangfeng facing peony pattern. In the book "*Huainanzi*" by Liu An in the Han Dynasty, the phoenix was called the bird of auspiciousness. Dragons and phoenixes are auspicious animals in people's hearts. Where there is a dragon, there is a phoenix. Peony is the king of flowers. The double phoenix peony means "the dragon and the phoenix are auspicious, the world is peaceful, and the rich and auspicious".

（三）海棠花、莲花等纹样/Chinese Flowering Crabapple, Lotus, and Other Patterns

海棠花与玉兰花、牡丹花、桂花相配在一起，取玉兰花的"玉"字、牡丹纹样"富足"、桂花纹样"贵重"，组成了"玉堂富贵"。这几种花都代表美好和富贵，组合在一起有富足有余的含义。莲花纹样是圣洁的代表，更是佛教神圣净洁的象征。莲花出尘离染，纯洁无瑕，是友谊的象征和使者，故而莲花纹样广泛应用于旗袍图案设计中，如图4-37所示。

Chinese flowering crabapple is matched with magnolia flower, peony flower and osmanthus flower to form the combination of "Yu Tang Fu Gui" which stands for a dignified, noble, and wealthy life. The lotus pattern is a symbol of holiness and Buddhist sacredness for its cleanliness and flawlessness, it also represents friendship. Therefore, the lotus patterns are widely used in the design of cheongsam patterns, as shown in Figure 4–37.

图4-37 玉兰、莲花纹样在旗袍中的应用
The Application of Magnolia and Lotus Patterns in Cheongsam

（四）缠枝纹/Tangled Branch Pattern

缠枝纹又名"万寿藤""转枝纹""连枝纹"。它是一种以藤蔓、卷草为基础提炼而成的传统吉祥纹饰。缠枝纹所表现的"缠枝"，以常青藤、扶芳藤、紫藤、金银花、爬山虎、凌霄、葡萄等藤蔓植物为原型。这些植物是吉祥花草，多为世人赞咏，缠枝纹就是这些藤蔓的

形象再现，常以植物的枝干或藤蔓为骨架，向上下、左右延伸。因为结构连绵不断，因此象征"生生不息"，寓意吉庆。

The tangled branch pattern is also known as "Wanshou vine" "Zhuanzhi pattern" and "Lianzhi pattern". It is a traditional auspicious decoration based on vines and curly grass. The "tangled branches" of the pattern took the vines of ivy, euonymus fortunei, wisteria sinensis, honeysuckle, creepers, campsis grandiflora, grape, etc.as creation prototypes. These plants are auspicious and mostly praised by people. The patterns created from them are often the branches or vines of the plant stretching outwards in different directions, and such continuous structure symbolizes "endless life" and auspiciousness.

缠枝纹是中国古代传统纹饰之一。除了美好的寓意，缠枝纹能够长远流传的一个原因在于其形态多变，可组成"缠枝莲""缠枝菊""缠枝牡丹"等图案。

The tangled branch pattern is one of the traditional patterns in ancient China. In addition to its good implied meaning, one of the reasons why it can be spread for a long time lies in its changeable shapes that can form patterns such as "twisted branch lotus", "twisted branch chrysanthemum", "twisted branch peony" and so on.

明代是中国吉祥文化发展的巅峰时期，缠枝纹作为这一时期广为流行的装饰纹样，以其回转缠绕的构成形式、丰富有序的组合法则，明确生动地表达出了明代百姓热爱生活、追求幸福的世俗心境。

The Ming Dynasty witnessed the greatest prosperity of Chinese auspicious culture. As a popular decorative pattern in this period, the twisted branch pattern clearly and vividly expresses people's love for life and worldly longing for happiness.

以象征富贵华丽的牡丹为主题的缠枝牡丹纹、以象征出世悠闲的菊花为主题的缠枝菊花纹、以象征外来文化的西番莲为主题的缠枝西番莲纹，这些纹样都反映出吉祥文化对缠枝纹样构图形式上的影响。缠枝纹是以藤蔓卷草经提炼变化而成，委婉多姿，富有动感，优美生动。因其结构连绵不断，在锦缎织物具有浓郁的装饰风格，与旗袍的造型和风格融合设计，更能体现其柔美的造型风格，在旗袍中广泛应用，如图4-38、图4-39所示。

The tangled peony pattern themed with the peony that symbolizes wealth and splendor, the tangled chrysanthemum pattern themed with the chrysanthemum that symbolizes leisure and easy minds, and the tangled passionflower pattern themed with passionflower that symbolizes foreign cultures are all the reflections of the impact of auspicious culture on the composition of the tangled branches patterns. Made by refining and changing the curly vines, this pattern is euphemistic and colorful,

图4-38　缠枝花纹在旗袍中的装饰
Decoration of Tangled Branche Pattern in Cheongsam

dynamic, beautiful, and vivid. Because of its continuous structure, it has a strong decorative style on the brocade fabric, which by integrating with the shape and style of the cheongsam cab better shows its soft style, thus being widely used in the cheongsam, as shown in Figure 4-38 and Figure 4-39.

图4-39　缠枝花纹在旗袍领口、袖口中的装饰
Decoration of the Tangled Branch Pattern on the Neckline and Cuffs of the Cheongsam

五、团花纹/Round Pattern

团花纹也称"宝相花""富贵花"，以精美细致、饱满华丽的艺术样式著称。团花纹外形圆润成团状，内以四季草植物、飞鸟虫鱼，吉祥文字，龙凤等纹样构成图案。其寓意是金玉满堂，荣华富贵。团花纹常见于袍服的胸、背、肩等部位。有放射、旋转、对称式等结构。

Round pattern, also known as "Baoxianghua" and "Fuguihua", is famous for its exquisite, meticulous, and gorgeous artistic style. The pattern is round in shape, composed of four plants, birds, insects and fish, auspicious characters, dragon and phoenix and other patterns on the side. It implies great prosperity and wealth, which is commonly seen on the chest, back and shoulders of robes. Its structures include the radiation, rotation, symmetry, and others.

团花纹样是喜庆吉祥的民俗的内涵体现。服饰中的吉祥图案有着漫长的发展过程，在这一过程中，材料工艺的变化更新，异域文化的注入和融合，本民族的审美意识等因素，都起到了相当大的作用，如图4-40所示。

Round pattern is a reflection of the festive and auspicious folklore connotation. The auspicious patterns in clothing have undergone a long development process. The changes and updates of materials and techniques, the injection and integration of exotic cultures, and the self-consciousness of the national aesthetic consciousness have all played a profound role in this process, as shown in Figure 4–40.

图4-40　团花纹在旗袍中的应用
The Application of the Round Pattern in Cheongsam

六、几何纹样/Geometric Patterns

菱形纹、格纹等几何图案，没有太多的象征意义，只是简单地作为一种图案排列，可以增加旗袍的美感。

Geometric patterns such as rhombus and plaid don't have much symbolic meaning, they are just simply arranged as a pattern to increase the beauty of the cheongsam.

几何纹是几何图案组成的有规律的纹饰，有龟甲、双距、方棋、双胜、盘绦、如意等形式。春秋战国时期，几何纹在其他纹饰衰退后成为主体纹饰。隋唐时期，纹样造型丰腴、主纹突出，地部疏朗，常用对称构图，色彩鲜丽明快。至五代纹样渐趋写实细腻。几何纹在宋代继续流行，并对明清时期的织锦产生了深刻的影响。

Geometric patterns are regular patterns composed of geometric patterns, covering tortoise shell, double distance, square chess, Shuangsheng (a pattern with two rhombus and circles intercrossed), strip, Ruyi, and so on. During the Spring and Autumn Period and the Warring States Period, geometric patterns became the main patterns as other patterns declined, and got further developed with the prominent main patterns, sparse ground, symmetrical structure, and bright colors during the Sui and Tang Dynasties. By the Five Dynasties, it gradually became more realistic and delicate. Its popularity lasted until the Song Dynasty, having a profound impact on the brocades in the Ming and Qing Dynasties.

民国旗袍与西方服饰审美在某些方面具有一致性。因此，旗袍的装饰图案也吸取了西方的特点，在受欢迎的几何图案中，有在面积区域分成的圆点纹样和方格菱形纹样，以及在线型区域分成的直线图形和非直线图形，如波浪纹，如图4-41、图4-42所示。

The aesthetics of Chinese cheongsam is consistent with the Western clothing culture to some extent. Therefore, the decorative pattern of the cheongsam also draws on the characteristics of the West. Among the popular geometric patterns, there are dot patterns and square rhombus patterns divided into areas, as well as straight lines and non-linear figures divided into linear areas, such as the wave pattern, as shown in Figure 4-41 and Figure 4-42.

图4-41　几何纹样在旗袍中的运用
Application of Geometric Patterns in Cheongsam

图4-42　几何纹样在旗袍中的创新运用
Innovative Application of Geometric Patterns in Cheongsam

（一）圆点纹样/Dot Pattern

圆点纹样简洁大方，颜色统一，相比较完全没有图案的旗袍，加上圆点纹样的旗袍，使穿着者多了一丝活泼青春的气息。

The dot pattern is simple and dignified, with uniform colors. Compared with the cheongsam without any pattern at all, the cheongsam with such a pattern makes the wearer look more energetic and young.

（二）方格菱形纹样/Checkered Diamond Pattern

方格菱形纹样在民国时期的应用颇为广泛，它打破了普通的线条，组合成常见的几何图形，通过颜色的渐变和线条的效果，拉长人体在视觉上的比例，而且菱形的不规则感让穿着者多了一丝成熟的感觉。

As a widely used pattern in the period of the Republic of China, checkered diamond pattern broke the ordinary feature of lines and combined them into common geometric figures, stretching the visual proportion of the human body through color gradients and lines. And the irregularity of the diamond shape made the cheongsam more mature.

（三）直线型纹样及变化线型纹样/Linear Pattern and Changing Line Pattern

直线型纹样分为横条和竖条，在普通的旗袍上加上均匀的线条使服装多了设计感。横条直线可以拉长人体的横向比例，使人看起来比较丰盈饱满。竖向直线则可以拉长人体竖向比例，让人看起来高挑轻盈。除了横竖向的直线，线条的形状可以是变化较多的波浪纹，也可以是弯曲的不规则纹路，每种变化的纹样给人的感觉都是不一样的。

The linear pattern is divided into horizontal strips and vertical strips, which. Adding an ordinary cheongsam with uniform lines makes the clothing look more fashionable. Horizontal lines can elongate the horizontal proportion of the human body, making people look plump, while vertical lines can stretch the vertical proportion of the human body, making the wearer look tall and light. In addition to these lines, the shape of the lines can also be wavy lines with more changes, or curved irregular lines. Each variation of the pattern brings different feeling of design and aesthetic.

第三节 丝绸旗袍装饰手法/Decorative Techniques of Silk Cheongsam

丝绸旗袍的装饰主要包括边饰、盘扣和刺绣，也涉及与之配套的其他服饰品，如首饰、鞋袜等。旗袍的装饰设计应突出穿着者的自然曲线，以简洁大方为美。旗袍的领口、袖口、下摆等装饰品种丰富，形式多样，并随造型变化。在丝绸旗袍的图案处理上，除面料本身的印花外，还可以结合传统刺绣，采用与流行趋势相结合的装饰手法，与面料一起烘托穿着者的个性特点。

The decorative techniques of silk cheongsam mainly consist of borders, knot buttons and embroidery, as well as other accessories such as jewelry, shoes and socks. The decorative design of the cheongsam should highlight the wearer's natural curve to show simplicity and elegance. The decorations on the neckline, cuffs, and hem vary in a wide range and change with the shape. In addition to the printed one on the fabric, the pattern of the silk cheongsam can be combined with traditional embroidery to make the wearer's personality prominent with the use of decorative techniques that follow fashion trends.

一、丝绸旗袍边饰与盘扣/Borders and Knot Buttons of Silk Cheongsam

（一）边饰/Borders

旗袍的边饰是指装饰旗袍领、袖、襟、裾边缘的花边，其结构复杂、工艺精湛、花样多重，达到了线条装饰艺术的巅峰。它的色彩、图案、宽窄、厚薄与旗袍的其他装饰配伍，共同构成了穿着时的多元变化，直接影响旗袍的层次感和整体效果。用丝绸制作的传统边饰主要包括"镶""嵌""绲""宕"四种工艺类型，见表4-1。它们之间可以根据款式需要灵活搭配，可以分为单色镶边、单色绲边、嵌边、镶嵌绲混合、三色镶边等。另外，蕾丝花边也常作为丝绸旗袍的装饰。

The borders of the cheongsam refer to the lace that decorates its collar, sleeves, front and hem, with complex structure and exquisite craftsmanship, reaching the pinnacle of line decoration art. Its color, pattern, width, thickness and other decorations of the cheongsam together constitute the diverse changes when wearing the cloth, therefore having direct implications for the layering and overall effect of the clothing. There are four primary types of traditional borders made of silk, namely, "braid（Xiang）", "panel（Qian）", "piping（Gun）" and " bias strip（Dang）", as shown in Table 4-1. They can be flexibly matched according to the needs of the style, and divided into single-color braid, single-color piping, panel, the mix of braid, panel, and piping, three-color braid, etc. Besides, lace is often used as the decoration of silk cheongsam.

表4-1 传统丝绸边饰工艺类型特征

镶	嵌	绲	宕
装饰在服装的表面，突出面料间的拼接效果，有一定宽度，可多条一起出现	常与镶边搭配，可分作埋线和不埋线两种做法	用织带，或将面料斜裁45°，加工成细条后，手工缝制在衣服的边缘	与"绲边"或"镶边"搭配出现，中间隔空一段距离形成平行线

Table 4-1　Characteristics of Traditional Silk Border Techniques

Braid	Panel	Piping	Bias Strip
It is decorated on the surface of the clothing with a certain width to highlight the splicing effect between the fabrics, and can appear together in multiple pieces	It is often used with braids, and can be divided into visible and invisible panels according to the process	It is the thin strip sewed by hand on the edge of the garment with webbing or cut from the fabric at a 45° angle	The lines matched with "piping" or "braid", with a distance in the middle to form the parallel lines

1.丝绸边饰/Silk Borders

（1）镶。"镶边"能够让整件旗袍的花型图案更加亮丽，根据穿着者的个性，采用与旗袍本身颜色相似的真丝绸缎裁剪成条状，把它"镶"在旗袍各个接缝处和边缘，突出面料图案的层次感，而不显单一。与其他边饰工艺相比，"镶边"通常会略宽一些，装饰部分更多在服装的表面，突出面料之间的拼接效果，如图4-43所示。

Braid. It is made by cutting the silk satin with a color similar to the cheongsam into strips according to the individual personality of the wear and is inlaid on the seams and edge of the cheongsam to highlight the fabric pattern, making the pattern more beautiful with layers. Compared with other borders, the "braid" is usually slightly wider, and more decorated on the surface of the garment, to underline the splicing effect between the fabrics, as shown in Figure 4-43.

图4-43　镶边旗袍
Cheongsam with Borders

（2）嵌。"镶边"的边上时常会带一条细线，被称作"嵌条"，可以让"镶边"更有立体感，是旗袍固有的特色，分为"外嵌"和"里嵌"。外嵌通常单独出现或与镶边搭配运用，里嵌通常和镶边或绲边搭配，也可以在两条镶边的中间，称作"一镶一嵌""两镶一嵌"等，工艺很灵活，搭配较随意，在旗袍上占的面积也不大，但是极费工时。"嵌条"又分作埋线和不埋线的做法，埋线需要考虑埋进去的线的缩水性，如图4-44所示。制作时根据镶边的颜色，再结合旗袍本身的颜色和花型图案的颜色，用特制的布料熨烫成1.5～2cm的细条状"嵌条"，用手工缝制在镶边和大身的面料边缘之间，如图4-45所示。

Panel. It is the thin line on the edge of the braid, making the braid more three-dimensional. As an inherent feature of the cheongsam, it is divided into the outer panel and the inner panel. The former is often used alone or in combination with the braid, while the latter is usually matched with the braid and piping, or in the middle of the two braids, including the one braid and one panel, two braids and one panel, etc. It features prominently in flexible craftsmanship and changeable match, only occupying a small area in the cheongsam, but taking a long time to make. In terms of its processing method, it can be divided into the thread-burying panel and the thread-non-burying one. For the former, the shrinkage of the thread buried should be considered, as shown in Figure 4-44. A special fabric is ironed in line with the color of the braid, the cheongsam, and the pattern into a thin strip of 1.5-2cm width, which then is sewed on the edge of the braid and the clothing, as shown in Figure 4-45.

（3）绲。"绲"字不常见，经常用"滚"字代替这一专用名称。"绲"字的本义是织带，手工缝制在衣服的边缘，可以收拢布边，使其不易脱线，且对薄型面料服装有较好的定型作用，是非常基础的边缘装饰。"绲边"除用织带外，常见的还有将面料斜裁45°角，加工成细

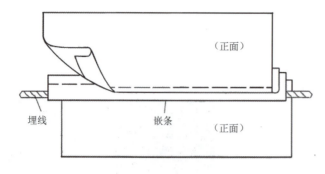

图4-44　嵌条中的埋线做法
Practice of Burying Threads in the Panel

图4-45　镶边和大身间的嵌条
Panel between the Braid Edge and the Clothing

条后使用，如图4-46所示。这个过程较为费料耗工，手工操作不易把握，斜裁面料便于旗袍包边时弯曲扭转，处理流畅后熨烫成形，常与镶边和嵌条工艺配合使用。

Piping. The word "绲" is not common in Chinese characters, thus is often replaced by "滚". Piping is a very basic edge decoration, originally meaning a webbing that is hand-sewn on the edge of the clothes to bunch it up. It also has a good shaping effect on thin fabric clothing. In addition to the webbing, it is common to cut the fabric at an angle of 45° and process it into thin strips for use, as shown in Figure 4-46. This process is material and labor-intensive, and the manual operation is difficult to grasp. Cutting fabric with the method of bias-cut makes it easier to bend and twist in piping which can get shaped by ironing. Therefore, such a border becomes a frequent match for braid.

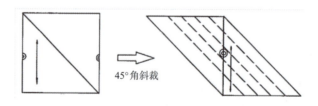

45°角斜裁

图4-46　裁绲条示意图
Diagram of Cutting Piping

（4）宕。"宕条"也经常作"荡条""档条"等，虽然看起来跟"镶"的工艺有点相似，但是"镶边"一般是一条挨着一条，而"宕条"是中间隔空一段距离再出现的，它一般与"绲边"或"镶边"一起出现，如同平行线一般，民国旗袍上常见的两条差不多宽细的平行线（一绲一宕），也称"电车轨"，如图4-47所示。一些老衣服看起来很像"两镶"或"两镶一嵌"的做法，仔细看就会发现中间那条是空的，用的就是"宕条"的做法。"宕"是用反差性极强的真丝单色绸缎裁剪成流线形或者波浪形，缝在领口、袖口、开襟、底摆等位置，使旗袍的装饰设计更富有张力，其制作更需耐心和熟练度，才可以将"宕条"准确地稳固在相应位置，使其看起来更平服，且与周围装饰相协调。"宕条"主要是配合单绲边、单镶边使用的工艺，以强化镶、绲效果，而且使用的质地一般都要与镶边、绲边一致，间距等于宕条宽，也有一色多道边或多色多道边的绲边与宕边相结合的工艺。

Bias Strip. It is also often referred to as "荡条", and "档条" in Chinese, which looks a bit similar to the braid. The difference between the braid and the bias strip lies in that the former is decorated one by one, while the latter has distance in between. It is usually accompanied by the piping or braid in the shape of parallel lines. The two parallel lines of about the same width (one curling and one dangling) commonly seen on the cheongsam of the Republic of China are also known as "the electrical-vehicle track", as shown in Figure 4-47. Some old clothes had something similar to the two braids or two braids with one panel. While, when looking closely, empty can be found in the middle, which was how the bias strip was applied. It is made of monochromatic silk satin with strong contrast that is cut into a streamlined or wavy shape and sewn on the neckline, cuffs, front, bottom hem, etc. of the cheongsam to make the clothing have greater tension. It requires profound patience and proficiency to fix the bias strip accurately on the corresponding position, to make it more flat and harmonious with the surrounding decoration. This border is mainly used in conjunction with the single piping and braid to strengthen the effect of the edge, and the texture used is generally consistent with the piping and braid, and the spacing is equal to the width of the strip. There is also a process of combining a single-color with a multi-lane edge or a multi-color with a multi-lane edge.

图4-47 民国旗袍上的一缲一宕
One Piping and One Bias Strip on the Cheongsam of the Republic of China

2. 蕾丝花边/Lace

手工蕾丝是欧洲传统的手工艺，非常精致美丽。工业革命后，机器生产代替了手工制作，蕾丝的使用被普及，化学工业在第二次工业革命时得到广泛发展，1910年，美国开始大规模生产人造纤维，各式蕾丝花边也大量生产。将蕾丝花边结合边饰工艺镶缝在旗袍边缘，不仅简化了旗袍烦琐的边饰制作过程，而且可以体现出女性的柔美及追求高雅精致的格调，具有时代特征，同时也拓展了中国女装的装饰手法。

Handmade lace is a traditional European handicraft, which is very delicate and beautiful. After the Industrial Revolution, machine production replaced manual production, leading to the popular application of lace. And as the chemical industry was widely developed during the Second Industrial

Revolution, the United States began to produce man-made fibers on a large scale in 1910, which boosted the mass production of various types of lace. Combining the lace and the above borders on the edge of the cheongsam not only simplified the cumbersome process of the cheongsam borders, but also highlighted the feminine beauty and the pursuit of elegant and refined style. The decorative method had the characteristics of the times, and expanded the possibility of Chinese women's clothing.

蕾丝花边多为黑色，也有白色和彩色的，宽窄不一。其中，宽蕾丝边的图案复杂，装饰在旗袍上效果突出，搭配素色面料更加抢眼。搭配鲜艳花色面料，大多用细窄的蕾丝花边，由于面料本身的图案很抢眼，搭配太夸张的蕾丝花边容易造成视觉上的杂乱感，如图4-48所示。

Lace is mostly black, but also white and colored, with different widths. Among them, the wide lace is characterized by complex patterns and splendid decorative effect, becoming especially eye-catching when matched with plain fabrics. The thin lace is usually used for fabrics with bright colors, since an exaggerated lace will lead to a poor visual effect of the fabric that's already eye-catching, as shown in Figure 4-48.

图4-49是北京服装学院服饰博物馆的馆藏旗袍，该旗袍不仅装饰了大红与粉红的双色镶边，而且装饰有与面料颜色接近的彩色蕾丝花边，与旗袍的烂花绒面料相得益彰，尽显旗袍本身的豪华气派。

As shown in Figure 4-49, it is the cheongsam in the collection of the Beijing Institute of Fashion Technology Clothing Museum, decorated with red and pink two-color borders and colored lace with similar color to the fabric, which brings out the best of velvet fabric, fully showing the luxurious style of the cheongsam.

图4-48　镶宽、窄蕾丝花边旗袍
Cheongsam with Wide and Narrow Lace

图4-49　北京服装学院服饰博物馆的镶蕾丝花边旗袍
The Cheongsam Inlaid with Lace at the Beijing Institute of Fashion Technology Clothing Museum

（二）盘扣 /Knot Button

盘扣是旗袍上重要的组成部分，承担着实用和装饰的双重作用。盘扣不仅样式丰富，而且制作工艺独具中国特色。盘扣可以分为直扣和花扣两种类型。直扣虽然结构简单，但是使用普遍，常见的有葡萄纽、一字纽等；花扣的式样多以各种花卉、植物、动物等的形状作为原型，还有以福、禄、寿、喜等字作为原型的，常见的有菊花纽、蜻蜓纽、寿字纽、如意纽、琵琶纽等。

As an important part of the silk cheongsam, the knot button plays a dual role of practicality and decoration, with abundant styles. Its production process is of unique Chinese characteristics. It can generally be divided into two types: straight knot button and flower knot button. The former is simple in structure but popular in application, including grape buttons, one-word buttons, etc. The latter is mostly based on the shapes of various flowers, plants, animals, etc., as well as the Chinese characters of 福, 禄, 寿, 喜, etc., including the chrysanthemum button, dragonfly button, longevity button, Ruyi button, Pipa button and so on.

盘扣是体现旗袍美感与个性的最好细节，它与领型、花纹及装饰细节搭配，显示了旗袍的价值和设计者的匠心巧思。图4-50是一件清代一字形金属扣服饰，由于镶嵌绳边非常复杂，盘扣不需要再有过多的装饰变化。如图4-51所示，这种复杂图案的盘扣在制作过程中需要将金属丝嵌入其中，以达到固定造型的目的。

The knot button is the best detail that reflects the beauty and personality of the cheongsam. Matched with the collar shape, pattern and decorative details, it will show the value of the cheongsam and the ingenuity of the designer. As shown in Figure 4-50, it is a piece of clothing with a linear shaped metal knot button in the Qing Dynasty. Because the inlaid piping is very complicated, the decoration of the button shall be as simple as possible. As shown in Figure 4-51, the complex pattern of the button needs to be embedded with metal threads in production to fix the shape.

图4-50　一字形金属扣
Linear Shaped Metal Knot Button

图4-51　复杂图案的盘扣
The Knot Button of Complex Pattern

盘扣的花型应考虑与面料的图案、颜色相呼应，做到和谐统一。复杂的花型装饰在颈部和胸前，视觉的重心上提，同时也提升了穿着者的气质。图4-52是一件专门在苏州定做的翠色真丝蝶恋花旗袍，衣身上的蝴蝶刺绣栩栩如生，与之相呼应的领口的盘扣，犹如从图案中跳出来伏在领口与胸前的蝴蝶。

The pattern of the knot button should be consistent with the pattern and color of the fabric to make the garment harmonious and unified. The complex flower pattern is decorated on the neckline and chest, which improves the visual effect and the wearer's temperament. As shown in Figure 4-52, this emerald silk cheongsam with butterfly and flower patterns was specially custom-made in Suzhou, and the butterfly embroidery on the clothing is lifelike, which, matched with the button on the neckline, is so vivid that seems to jump out of the garment.

图4-52　盘扣的形状色彩与面料图案相呼应
The Shape and Color of the Knot Button is Harmonious with the Fabric Pattern

二、苏绣与丝绸旗袍/Su Embroidery and Silk Cheongsam

丝绸旗袍图案的实现手段常以刺绣或面料的提花织造为主。在旗袍造型变化少、以直身为主的早期，为了更好地达到旗袍的适应性，或出席隆重场合时穿着，人们花了更多的心思在图案上，以显示其贵重程度。苏绣是以苏州一带为生产中心的江苏刺绣，因其拥有图案秀丽、色彩典雅、针法丰富、绣工精细的地方风格，而名列四大名绣之首，其工整、细腻、精致的传统艺术特色被南北方的刺绣所借鉴。苏绣运用在丝绸旗袍上，能将吴门画派的手绘艺术、书画之美融入服装，透出一种优雅韵味，并做到图案细节与旗袍的款式、面料及穿着者的身份气质相呼应，形成独具地方特色的装饰风格，如图4-53所示。

The silk cheongsam patterns are often based on embroidery or jacquard weaving. Especially in the early days when the shapes of cheongsam were limited and the clothing body was straight, to make the cheongsam more adaptive, or to serve for official wearing, people paid greater attention to the pattern to show its preciousness. Su Embroidery is Jiangsu embroidery with Suzhou as its production centers. It ranks first among the four famous embroidery because of its local style of beautiful patterns, elegant colors, abundant stitching methods, and fine embroidery workmanship. The neat, delicate traditional art features of Su Embroidery have been referred to by embroidery in the north and south. Applying Su Embroidery to silk cheongsam can integrate the hand-painted art and the beauty of calligraphy and painting of Wumen School into clothing, thereby showing impressive elegance, and achieving a harmonious relation between the pattern details with the style, fabric and identity of the wearer, forming a unique decorative style with local characteristics, as shown in Figure 4-53.

图4-53　苏绣与丝绸旗袍
Su Embroidery and Silk Cheongsam

　　与文人画绣相比，苏绣旗袍在题材的选择上更简单抽象，写意夸张，针法上更具规律性，用线纹理明显，用色善用对比色，注重色彩间的呼应关系，质朴和谐。苏绣主要通过对不同针迹纹理和针法的变换运用，来表现色彩的晕染过渡；它还采用协调与对比、隐含与突出这种中国式的色彩与技法搭配，来表现新的意蕴和独特的审美价值，进而实现雅俗共赏的审美特性。"苏意"代表着雅致的生活态度、精致含蓄的审美取向和时尚的文化品位，苏绣旗袍中细腻精巧的绣法正是对"苏意"精髓的吸收与运用。苏绣旗袍是"苏意"文化背景熏陶下的产物，也是"苏意"文化的一种表现形态与载体。

Compared with literati painting and embroidery, Su Embroidery cheongsam prefers simple and abstract subjects, which are exaggerated in freehand brushwork, and regular in stitching, with obvious thread texture, good use of contrasting colors, and emphasis on the relationship between colors, looking simple and harmonious. It intends to express the transition of colors through the transformation and application of different stitches and texture, as well as the new meaning and unique aesthetic value by adopting coordination and contrast, implicit and prominent Chinese-style color and technique match, thereby making it favored by both the noble and the ordinary. "Su Yi (the style of Suzhou)" represents an elegant attitude of life, refined and subtle aesthetic orientation, and fashionable cultural taste. The delicate embroidery method in Su Embroidery cheongsam reflects the absorption and application of such thought of life. Su Embroidery cheongsam came into being in the cultural background of "Su Yi", and it is also a form and carrier of "Su Yi" culture.

三、丝绸旗袍整体搭配/Matching of Silk Cheongsam

穿着丝绸旗袍时，一般佩戴小巧精致的耳坠、珍珠项链、手镯、宝石纽扣等首饰，足蹬各种绣花鞋，这种搭配展现了地道的中式风格。随着社会的发展，女性将西式外套、丝袜、手表、高跟鞋、丝巾、坤包等作为旗袍的搭配，这也是典型和直观的中西合璧的搭配方式。

When wearing a silk cheongsam, One usually wear small and delicate earrings, pearl necklaces, fine bracelets, gem buttons and other jewelry, paired with different embroidered shoes, which show an authentic local charm. With the development of society, women have taken Western–style jackets, stockings, watches, high–heel shoes, silk scarves, bags, etc. to match with the cheongsam, which becomes a typical and intuitive way to integrate the Chinese and Western clothing style.

丝绸旗袍的可搭性极强，称得上是"百搭"的服饰。天凉时外加短背心、针织衫或披肩，寒冷的冬季搭配西装外套、毛呢大衣，在领、袖处加以人造毛皮装饰，更是一种时髦摩登的穿法。丝绸旗袍与时尚物品的搭配可以提升旗袍的整体效果，衬托人物或秀气自然，或成熟婉约，或温婉大气等不同的性格特点，通过旗袍的不同搭配，可以很好地辨识穿着者的身份性格等内在气质。百变的搭配也能使丝绸旗袍的功能最大化，以满足各种穿着场合的需求。

Due to its strong adaptability, the silk cheongsam is regarded as a "universal" dress. Adding a short vest, knitted sweater or shawl when the weather is cool, and matching it with a suit jacket or woolen coat, and decorating the collar and sleeves with faux fur in the cold winter is fashionable and modern. The combination of silk cheongsam and fashionable accessories can greatly improve the overall effect of the cheongsam and set off different characteristics such as natural delicacy, maturity and gracefulness, and gentleness. These can be well identified through the different matching of cheongsam. The variety of matching can also maximize the function of silk cheongsam and adapt to various occasions.

思考题

（1）概述丝绸旗袍面料的分类与特点。

（2）论述丝绸旗袍的整体搭配如何做到因人而异。

Questions

（1）Summarize the classifications and characteristics of cheongsam silk fabrics.

（2）Talk about how to customize the silk cheongsam individually.

参考文献
References

［1］袁杰英．中国旗袍［M］.沈蓁，译.北京：中国纺织出版社，2000.

［2］包铭新．近代中国女装实录［M］.上海：东华大学出版社，2004.

［3］卞向阳.论旗袍的流行起源［J］.装饰，2003（11）：68-69.

［4］陆洁，余祖慧．由英译名解析旗袍缘起过程中的多元融合［J］.现代丝绸科学与技术，2020（5）：15-17.

［5］春梅狐狸．图解中国传统服饰［M］.南京：江苏凤凰科学技术出版社，2019.

［6］陆洁．从内在需求看民国女性穿着方式的嬗变［J］.丝绸，2020（2）：91-96.

［7］朱博伟，刘瑞璞．旗袍三个发展时期的结构断代考据［J］.纺织学报，2017（5）：115-121.

［8］刘瑞璞，刘维和.女装纸样设计原理与技巧［M］.北京：中国纺织出版社，1991.

［9］陈礼玲．旗袍结构设计和工艺演变研究［D］.无锡：江南大学，2010.

［10］俞跃，赵明．中国传统旗袍工艺与造型的关联性探析［J］.丝绸，2013（5）：23-27.

［11］张浩，郑嵘，徐枫，等．旗袍空间省量分配与侧缝形态关系［J］.北京服装学院学报，2006（3）：39-45.

［12］陈美仙.论旗袍的结构与制作工艺流程［J］.上海视觉，2017（2）：77-87.

［13］郑嵘，张浩.旗袍传统工艺与现代设计［M］.北京：中国纺织出版社，2000.

［14］陆洁．传统旗袍的手缝边饰工艺与设计技法［J］.毛纺科技，2021（1）：61-65.

［15］张丹烁.民国旗袍的装饰研究及现代设计创新［D］.北京：北京服装学院，2012.

［16］陆洁．当代高级定制服装中运用织染工艺的方法和规律［J］.毛纺科技，2019（3）：52-56.

［17］加佳，张竞琼.近代女袄中的苏绣针法与配色探析［J］.丝绸，2014（3）：48-59.